D1196545

Applied Multivariate Statistics
in Geohydrology
and Related Sciences

Springer

Berlin
Heidelberg
New York
Barcelona
Budapest
Hong Kong
London
Milan
Paris
Santa Clara
Singapore
Tokyo

Charles E. Brown

Applied Multivariate Statistics in Geohydrology and Related Sciences

With 72 Figures and 64 Tables

Springer

Dr. Charles E. Brown
International Geohydroscience and
Energy Research (IGER) Institute
15094 Wetherburn Drive
Centreville, Virginia 20120 USA

and

Department of Chemistry
George Mason University
Fairfax, Virginia 22030-444 USA

E-mail: iger.inst@mci.com

ISBN 3-540-61827-9 Springer-Verlag Berlin Heidelberg New York

Library of Congress Cataloging-in-Publication Data

Brown, Charles E.
Applied multivariate statistics in geohydrology and related sciences / Charles E. Brown.
p. cm   Includes bibliographical references (p. –) and index.
ISBN 3-540-61827-9 (hardcover : alk. paper)
1. Hydrogeology – Statistical methods. 2. Multivariate analysis.
I. Title.
GB 1001.72.S7B76   1998

© Springer-Verlag Berlin Heidelberg 1998
Printed in Germany

The use of general descriptive names, registered names, trademarks, etc. in this publication does not imply, even in the absence of a specific statement, that such names are exempt from the relevant protective laws and regulations and therefore free for general use.

Production: ProduServ GmbH Verlagsservice, Berlin
Typesetting: Fotosatz-Service Köhler OHG, Würzburg
Cover design: Design & Production, Heidelberg
SPIN: 10534514   32/3020-5 4 3 2 1 0 - Printed on acid-free paper

Preface

It has been evident from many years of research work in the geohydrologic sciences that a summary of relevant past work, present work, and needed future work in multivariate statistics with geohydrologic applications is not only desirable, but is necessary.

This book is intended to serve a broad scientific audience, but more specifically is geared toward scientists doing studies in geohydrology and related geosciences. Its objective is to address both introductory and advanced concepts and applications of the multivariate procedures in use today. Some of the procedures are classical in scope but others are on the forefront of statistical science and have received limited use in geohydrology or related sciences.

The past three decades have seen a significant jump in the application of new research methodologies that focus on analyzing large databases. With more general applications being developed by statisticians in various disciplines, multivariate quantitative procedures are evolving for better scientific application at a rapid rate and now provide for quick and informative analyses of large datasets. The procedures include a family of statistical research methods that are alternatively called "multivariate analysis" or "multivariate statistical methods".

This book addresses specific papers and topics that cover a broad spectrum of applications of multivariate quantitative procedures (or multivariate statistical techniques) in geohydrology and related sciences. Recently, there has been a growing and greater interest among geoscientists, especially geohydrologists, in using multivariate quantitative procedures with applications that are parsimonious. Within most of the applications, the methods have brought about a more quantitative approach to the interpretation and analysis of geohydrologic and other data that are routinely compiled into very large databases.

One main purpose of this publication is to focus attention on multivariate quantitative procedures that have found unique applications in geohydrology. It is hoped that this will not only serve to stimulate further interest in these mathematical procedures, but will also encourage others to apply and enhance these procedures when they are appropriate for their own data.

Part 1 includes Chapter 1 and 2 which discuss the introductory techniques and background of multivariate statistical procedures.

Part 2 covers variable-directed techniques requiring normality of data and includes Chapters 3 to 8 which cover topics and numerical examples on correlation, canonical correlation, multiple regression, and multivariate analysis of variance and covariance.

Part 3 covers variable-directed techniques which do not require normality of data and includes Chapter 9 which covers topics and numerical examples on principal components.

Part 4 focuses on individual-directed techniques that require normality of data and includes Chapter 10 which covers topics and numerical examples on discriminant analysis.

Part 5 also focuses on individual-directed techniques that do not require normality in the data and includes Chapters 11 and 12 which cover topics and numerical examples on cluster analysis and logistic regression.

Part 6 includes various other multivariate methods for exploring data and includes Chapters 13 to 15 which cover coefficient of variation, correspondence analysis, and multivariate probit analysis.

Part 7 includes multivariate measures of time and space and includes Chapters 16 and 17. Topics include multivariate time series analyses and multivariate spatial statistics.

Part 8 includes topics, examples, and concepts on graphical analyses that enhance the understanding of the theory and applications of the procedures. It includes Chapters 18 and 19 that cover graphical concepts and summary conclusions. Chapter 19 is the summary of methodologies. An appendix covering some of the mathematical concepts is also provided for quick reference and definitions.

This book is different from other texts on the subject because its focus is on applications in the environmental and geological sciences and especially geohydrology, and for this reason I have chosen a number of significant papers from geohydrology, related earth sciences, and biology to reference. Other texts that are in publication usually provide the programming aspects of the various methods, but this is not done in this volume because the focus is directed toward application. The necessary computer programming aspects with pertinent discussions can be found in all the user manuals of popular computer software packages covering statistical methods (Statistical Package for the Social Sciences-SPSS; Statistical Analysis System Institute-SAS, Biomedical Computer Programs-BMDP; and Minitab, to mention a few). The author has used all of these software packages in past work and highly recommends that the choice of which to use is left to the user in accordance with their computing power. The rapid rate of change in user documentation demands that we do not duplicate the state-of-the art programming that is provided elsewhere.

This book – Applied Multivariate Statistics in Geohydrology and Related Sciences – is also a comprehensive compilation of significant investigations and discussions that synergistically apply a wide array of mainstream quantitative procedures. These procedures are truly quantitative procedures because they engage hypothesis testing and criterion testing, and require that rigorous mathematically-based application rules are followed. Most or all of the methods have evolved to an advanced level of application and the performance of the applied test can be exactly specified and demonstrated. The use of criteria testing in multivariate analysis is where most reward is achieved and separates a qualitative from a quantitative application. This volume will serve as a guide to enhance understanding of these mathematical procedures that are readily embraced and used by nonscientists, as well as statisticians and researchers in the sciences.

Because databases in the environmental and hydrological sciences continue to grow at an alarming rate, new methods must be developed and new pathways explored for analyzing the accumulated information. We can focus on future needs in this scientific arena by carefully analyzing past work examples. The discussions in this book are meant to enhance that effort and to be an introduction and guide that assimilates some of the vast knowledge needed to correctly use the techniques, graph the data, and perform criterion or significance testing during the analyses. The analyses in these discussions adhere to the principle of parsimony and "Occam's Razor", i. e., "that it is vain to do with more what can be done with less". The topics are covered in a comprehensive manner to improve scientific exposure and application on the international fronts.

Multivariate quantitative procedures have been widely used for analysis of large datasets in other disciplines such as psychology, education, and especially biology, so much so, that procedures are growing in importance in these fields at a very rapid rate and it is necessary that we occasionally explore reference papers in these fields. However, the difference in application between science and education or psychology is that variables measured in the physical sciences are objective and physical in nature, allowing for less subjectivity in the final analysis.

The methods are expected to continue to grow and evolve in parallel with the growth of new expansive databases and the increasing computational capability of computers. This subject area in most cases may be non-traditional in the geosciences, but these procedures are taught in classes at most universities in either science, biology, or statistics departments and are important additions to understanding new research findings in the sciences.

The two major divisions of this book are based on whether the techniques are sample-directed procedures or variable-directed procedures. This book is divided into four minor subdivisions that are based on procedures that require data that assume multivariate normality (Parts 2 and 4) and procedures that do not assume multivariate normality of data (Parts 3 and 5).

In Part 1, which includes the Introduction, I define more fully the overall rationale for selecting and organizing the topics that are included in this volume. The attempt here is to integrate univariate, bivariate, and multivariate statistics through a short introduction of the main concepts and brief definitions. The theme of this book is centered on better data analysis and quantification in geohydrology and other related sciences. The topics are chosen that best convey a common and appropriate use of the methods, that best describe the approach, and that best illustrate the synergism of methods in geoscientific analysis. For this reason, I have on occasion crossed over into other disciplines, especially biostatistics, to find significant topics for inclusion.

I remain optimistic and hope that this book will help to expand and provide the necessary knowledge that is required to define the broad capabilities of multivariate procedures so that they can be applied in new and more diverse ways to solve geoscientific problems, i. e., through parsimonious applications. The author believes that true advancements in this area will come only if there is very wide exposure and use. In a book such as this, which is published in an age of exponentially accelerating growth of scientific literature, there will always be the unavoidable omission of some very good reference papers and subjects that should possibly be included because they are unique and difficult to access. However, I regret such omissions and extend sincere thanks to readers and authors perservering in these areas of research. This book attempts to fill the void when papers that discuss significant multivariate topics, such as those usually found in journals and other publications, are not accessible.

Charles E. Brown

Acknowledgements

Although the selection of referenced papers and discussions in this volume is based solely on my reasoning, the work of many scientists is the backbone of this book and includes information and experience gained from many classes taken and instructed at various institutions of higher learning such as the Pennsylvania State University, George Mason University, Virginia State University, and the Institute of Professional Education. I have received invaluable suggestions from scientists from various disciplines inside and outside the geosciences arena. I am especially indebted to the late John C. Griffiths, Professor Emeritus, and Professor Richard R. Parizek, both of the Pennsylvania State University, for providing the opportunity to purse graduate study in this area of the geosciences. I am very much appreciative of the insight gained from class lectures of the late Dr. John C. Griffiths, Dr. John J. Miller (George Mason University), and Dr. Dallas Johnson (Kansas State University and the Institute of Professional Education). I have also benefitted from students with whom I have worked and taught. I thank the late Professor Mack Gipson (Virginia State University and University of South Carolina) for my introduction to study in the geological sciences. This work is dedicated to the late Professor Gipson and late Professor Griffiths for providing invaluable guidance in many geoscientific endeavors.

To my colleague and friend, Reginal Spiller, hydrogeologist, I thank you for your support, suggestions, and review.

To my wife, Sadie, and daughters, Karen and Carla, I am especially indebted for support and encouragement throughout this effort, and I could not have completed this work without you.

To my mother, Gretchen, and father, Warren, and the rest of my family, thank you for encouraging me to study rocks instead of throwing them.

To the dedicated editorial staff of Springer-Verlag, I am especially thankful for their comprehensive review of this manuscript, and the careful guidance given throughout the report preparation process.

To the readers, thank you for promptly reporting any errors or concerns to the author, and I take full responsibility for all errors and omissions.

Contents

Introduction to General Statistical and Multivariate Concepts

CHAPTER 1

General Concepts

1.1
Concepts

Statistics is that branch of mathematics which deals with the analysis of data, and is divided into descriptive statistics and inferential statistics (statistical inference). Multivariate statistics is an extension of univariate (one variable) or bivariate (two variables) statistics. It allows a single test instead of many different univariate and bivariate test when a large number of variables are being investigated.

1.2
Definitions

Population. The complete collection of all items being studied.

Samples. A group of items chosen from a population.

Frequency. The number of observations in a class interval or category; a diagram of the frequency by class is a frequency distribution.

Mean. Same as average and is equal to the sum of the list of numbers in a list divided by the number of objects.

Standard Deviation. The square root of variance (equation of dispersion).

Variance. A measure of dispersion for a list of numbers; it is symbolized by sigma squared.

Percentile. The pth percentile in a list such that p percent of the numbers are below it; for first quartile (25th percentile), 25% of the numbers are below this number.

Probability. The study of random or chance phenomena.

Hypothesis Testing. A procedure that involves collecting data and then making a decision as to whether a particular hypothesis (conclusion) should be accepted or rejected based on data analysis.

Null Hypothesis. The hypothesis that is being tested; the hypothesis indicating that the null hypothesis is wrong is the alternative hypothesis.

Type 1 Error. Denotes that the null hypothesis is false when it is actually true; the common procedure is to design a test so that the chance of committing a type 1 error is less than a specified amount (i. e., level of significance which is often 5 %).

Type 2 Error. Saying that the hypothesis is true when it is actually false.
Random Variable. A variable whose value depends on the outcome of a random experiment such as a sampling experiment.

1.3
Introduction to Statistical Concepts

1.3.1
Meaning of Statistics

It is important to point out initially that there is more than one analytical statistical strategy that is appropiate for analyzing most data. The choice of the techniques used depends on the intended audience, interrelationships of variables, investigator's preference, and application of the principle of parsimony, wherein simplicity in interpretation is of primary concern.

In this chapter, we shall discuss statistical terms that are important in introducing the subject of general statistics. This chapter however does not treat the subject comprehensively; it and preceding chapters are viewed as a foundation to acquaint the reader with basic concepts of statistics that underlie multivariate statistical procedures and techniques. It is important to realize that multivariate statistics are an extension of univariate and bivariate statistics and this allows us to perform a single analysis instead of many univariate and bivariate analyses when we have measured many variables. Multivariate statistics then represents the complete or more simply the general case, and univariate and bivariate analyses are special cases of the general multivariate model (Tabachnick and Fidell 1989). This concept will be discussed further in the next sections.

The valid use of any statistical techniques depends on a set of fundamental asumptions, whether related to type of data, size of data set, variation in data, or other differences. I will attempt to state these assumptions in a general manner as we proceed through the subject matter.

The word "statistics" has two different and contrasting meanings: (1) a collection of numerical data, or (2) the branch of mathematics that deals with the analysis of statistical data. The subject can be further subdivided into: (1) descriptive statistics, or (2) inferential statistics. Descriptive statistics is the process of obtaining meaningful information such as the average, mean, mode, or measures of variance from sets of numbers that are often too large to deal with directly. Inferential statistics or statistical inference is the process of using observations from a sample to estimate the properties of a larger population. In essence, inferential statistics is a method of inductive reasoning. The term population refers to all the things or objects in the group that are being studied and the sample is the group of items chosen from the population of interest. Because it is sometimes expensive, difficult, or impossible to survey an entire population, a sample is usually selected for analysis. To obtain an accurate analysis or to prevent inaccurate predictions, it is essential that a random and representative sample of the population be chosen. The true meaning of random or representative sample will be discussed in several later sections.

In order to fully comprehend statistical inference, it is a prerequisite to understand some concepts in probability. Probability and statistics are closely aligned and related. The questions posed by the contrasting applications are opposing. In probability, it is known how a process works and the task is to predict the outcomes of the process. In statistics, the process is not known, but we are able to observe the outcomes of the process. The nature of statistical inference is to use the information about the outcomes to address the nature of the process.

1.3.2
Probability and Randomness

Probability is described conceptually by observing some common and simple concepts such as flipping a coin, tossing a die, or drawing cards in an"unbiased" manner. For example, suppose you have 40 blue rocks and 60 red rocks in a sample collection, and you draw 10 "randomly" selected rocks from the box. By using probability theory, the expected outcome of your process calculated exactly is that four will be blue and six will be red. When we do not know probabilities in advance, we must use statistical inference to estimate them, and this process includes such techniques as estimating the mean, correlation, or other population parameters. The technique of sample to population reasoning is called "inductive" or inferential reasoning, whereas reasoning from population to samples is called deductive reasoning. Constraints in sampling are reflective of the target population being studied. Target populations are defined in three different ways: (1) hypothetical, (2) existent, or (3) available. Hypothetical populations, for example, represent the total volume of material occurring at some initial point in geological time. The existent population represents that part of the population presently existing, but perhaps, not all the population can be sampled because it is buried or is no longer intact. The available population represents that volume of material that is accessible for sampling. Sampling programs should thus be arranged to adequately represent the available population which requires that a sample be a 'statistically' random sample of the population wherein each item or sample has an equal chance or probability of being selected. The process of "randomness" can only be assured through adequate predefined procedures leading to sound quality control. The simple statistical concepts defined here are the basis of many multivariate procedures described in this text and should be understood in general principle.

1.4
Meaning of Multivariate Analysis

All parametric statistical methods: (1) univariate, (2) bivariate, and (3) multivariate are special applications of the general linear model (GLM) that will be discussed later in the text. Multivariate data results from measuring several different variables on each experimental unit or sampled object. Multivariate analyses combine variables to do useful work. The combination of the variables that is formed is based on the relations between the variables and the goals of the analysis, but in all cases, it is a linear combination of variables. A linear combination is defined as one in which variables are assigned weights, and then the products of weight and variable scores are summed to produce a score on a combined variable (Tabachnick and Fidell 1989). A variable is a characteristic of a population that can have different values. Almost all data are multivariate in scope and thus occur in all branches of science. The multivariate methods are all based on the techniques of matrix algebra (see Appendix) and this supporting mathematics prevents widespread miscommunication in terms of what methods can do. Multivariate data can be discrete (categorical) as well as continuous, but this text will be more concerned with continuous data.

Simply speaking, multivariate analyses are often concerned with finding relationships between both response variables and experimental units or samples. The analysis of discrete data is usually studied under the title of 'categorical data analysis'and will not be discussed to any length in this text. Multivariate analysis is based and usually operates on the analysis of one or more of the following mathematically defined concepts: "$p(p-1)/2$" different covariances among the variates, "p" means, and/or "p" variances. It is expected that the three types of statistics described above, i. e., means (averages), variances (spread or square of standard deviation) and covariances (measure of association or dependency), are the natural parameters of multivariate normal distributions and thus the focus of most techniques that are applied. The first two statistical concepts, i.e., means and variances, are also the focus of univariate statistical techniques and the extrapolation to multivariate techniques is easy when the third concept, covariance, is included and emphasized. This analogy is valid because to summarize multivariate distributions, we need to find not only the mean and variance of each of the "p" variables, but also the correlations or covariances between the "p" variables. The analyses are done on several different data matrices including the following: (1) correlation matrix, (2) variance-covariance matrix, (3) sum-of-squares and cross-products matrix, and (4) residuals matrix that includes a measure of the error of prediction.

One of the key requirements of multivariate statistics involves defining the relationships among the variables under study. When the association is assessed between any two variables, and the variables are characterized as either independent or dependent, the technique used is bivariate correlation. On the other hand, the prediction of a dependent variable from an independent variable is done through bivariate regression.

Multiple correlation assesses the relationship between a dependent continuous variable and a set of continuous independent variables. The independent variables may be combined to form a composite variable, and the multiple correlation between the dependent variable and the new composite variable is the multiple correlation. Likewise, multiple regression is used to predict the score (values) of a dependent variable from a group of independent scores. When the importance of the independent variables is prioritized in some ordered fashion, then the method used is a form of hierarchical multiple regression.

In canonical correlation, another important method, we are asked to determine the relationship between two sets of continuous variables, one set being independent and the second set being dependent, or we are asked to relate the predictive ability of one set from another set.

Multiway frequency analysis is used to define how discrete variables relate in general, and has two main classifications. If one looks at a dependent discrete variable and its relationship to a set of independent discrete variables, the method applied is "logit" analysis. In other cases, a dependent variable may not be involved at all.

The simple t-test is used in testing the difference between two population means based on differencies found between sample means, and analysis of variance (AOV) is used to test the null hypothesis of no difference between two or more population means as represented by their respective sample means. It is called single factor AOV with one independent variable and an independent variable that may also have different levels. The independent variable may be described as fixed or random, can have different levels, and may have repeated measures as part of the designs. Large F values, associated with a small probability of occurrence, define the test of the null hypothesis, and relate the ratio of the variance within groups to the variance between groups.

The analysis of covariance is a way of extracting pretreatment differences (nonrandom or undefined differences) of a dependent variable before comparing means of groups. We call the control variable a covariate and it is used to describe pretreatment variations (differences) or ratings. The criterion variable is the one that is considered to be the posttreatment variable. Hence the techniques involve pretreatment means, posttreatment means, and differences among groups based on testing posttreatment means using an F-test.

Multivariate analysis of variance (MANOVA) can include an analysis of variance design to include one or more independent variables (and multiple groups or levels of independent variable) and multiple dependent variables. MANOVA can evaluate differences amongst centroids for a set of dependent variables when there are two or more levels of independent variables. The "T^2" test is a special case of MANOVA for two groups and more than one dependent variable. Factorial MANOVA is an extension of MANOVA to statistical designs with more than one idependent variable and multiple dependent variables. Another special form of MANOVA is called "profile" analysis.

Multivariate analysis of covariance (MANCOVA) is a method used to analyze multiple dependent variables, with multiple covariates through an analysis of variance design. Factorial MANCOVA incorporates one or more covariates into a factorial MANOVA design.

Multiple discriminant analysis is a method that uses a set of clustering or discriminator variables as a discrimination tool for separation of individuals or samples into classes.

Factor analysis is based on an underlying statistical model for the data and is used to develop a set of theoretical, uncorrelated variables called factors that describe the variance in the original variables more parsimoniously. It assumes a statistical structure for the data.

Principal components is a method that is also used to develop a system of simplifying variables, but assumes no underlying statistical structure for the data. The technique asks the question of how the variables relate to one another or more simply, how do the variables cluster together.

These topics that are generally defined above will be treated more fully in other sections throughout the text. The brief definitions and discussions are very important in light of the complexity of analyses that are forthcoming. When priority is placed on importance of variables, the descriptor 'hierarchical' may be added to the multivariate method.

1.5
Univariate to Multivariate Mathematical Generalizations

During the process of collecting water or rock samples from a sampling area or site and measuring a set of properties for each sampled object, the measurements or set of measurements will define a vector,$[X] = [X_1, X_2, ,,, X_m]$, where there are 'm' measured characteristics or variables (Davis 1973). If samples, represented by vectors [X], are randomly selected from a population which is the result of many independently acting processes, the observed vectors will tend to be multivariate-normally distributed. A vector of random variables [X] is said to have a multivariate normal distribution if A'[X] has a univariate normal distribution for every vector of constants [A]. Futhermore, considered individually, each variate is normally distributed and characterized by a mean, μ_k and a variance, $\rho(k)^2$. The joint probability distribution is then a p-dimensional equivalent of the normal distribution, having a vector mean $[\mu] = [\mu_1, \mu_2, \mu_m]$ and variance generalized into the form of a diagonal matrix.

The covariance matrix, (Σ^2) has covariances, cov(ij), which occupy all of the off-diagonal positions of the matrix, and variances along the diagonal, so it is a matrix of variances and covariances (Davis 1973). If we consider a simple t-test of the probability that a given random sample of n-observations has come from a normal distribution with a specified mean, μ_0, and an unknown variance, the test equates to:

$$t = (\text{mean } X - \mu_0(n^{1/2})/(s^2)^{1/2} \tag{1.1}$$

An obvious generalization to the multivariate case is to substitute a vector of sample means for mean X and a vector of population means for μ_0, and a variance-covariance matrix for s-squared (square root). We can now study a vector of population means, μ, and a vector of sample mean X's. Similarly the matrix of variances and covariances and s^2 may be designated the matrix of sample variances and covariances. Both [mean X] and [μ] are commonly taken as column vectors, but they may be row vectors as well. We may substitute these vectors to get a vector of differences between the sample means and population means and substitute into the above equation (1.1). Further substituting and multiplying by transposes, we get single numbers to work with leading finally to a T-squared test, with proper substitutions. Thus we can test differences, equality of two vector means, equality of variance-covariance matrices, and other multivariate analyses using proper procedures.

A linear combination is the combining of variables so that variables are assigned weights and then products of weight and variable scores are summed to produce a score on a combined variable. Different combinations can be formed but the combinations will usually always be linear or nearly so.

Strategies for dealing with overlapping variance can be complex. Overlapping variance is a function of correlations between variables, and the outcome of analyses is different based on the order of entry or deletion of variables into the analyses. Variables which are correlated with each other are nonorthogonal where as variables that are not correlated are orthogonal. A standard versus a hierarchical analysis treats overlapping variance in different ways. In a standard analysis, the overlapping variance contributes to the size of the summary statistics, but is not assigned to either variable and is disregarded in assessing the contribution of the variable to the solution. Some variance is not assigned to either variable during a standard analysis. For a hierarchical analysis, the procedure assigns the variance in the relationship between variables on the order of entry, with the first being assigned both individual or unique variance and any overlapping variance it has with other variables. Hence the relative contribution to the variance by variables is entry order dependent, with low-priority variables being assigned their unique variance plus any remaining overlapping variance. The researcher might choose to study the high priority variables first, then remove these before low-priority variables are studied. It is vitally important to understand that solutions change when different strategies for independent variable entry are chosen, thereby giving answers to entirely different questions and giving different prediction capability. Further, if there are too many variables in the analysis for the sample size, we run into a problem called "overfitting" and should resolve this problem by limiting the number of variables that are used. The rule of parsimony applies at this juncture of the analysis. Sometimes, the best compromise regression between a forward selection statistical regression (adding one at a time) and backward deletion statistical regression (include all and then delete one at a time) is the stepwise statistical regression that will be covered with an example later in this book. The procedures just described form the three versions of statistical regression wherein statistical criteria control the order of entry. Setwise regression is also used in some analyses (see Tabachnick and Fidell 1989).

1.6
Nature of References

The papers referenced and topics presented in this book deal mostly with physical and chemical geohydrology but also relate to other types of data analysis. Examples of unique applications are drawn from hydrologic journals and other water resources publications that report investigations in surface water hydrology, ground-water hydrology, water and soil geochemistry, and biostatistics. The application of multivariate quantitative procedures is not limited to geoscience data and work in other disciplines should be consulted as the applications evolve. Multivariate statistics is widely applied to problems in biology and soil science and some applications will be introduced later in this text.

Most of the published papers cover such general multivariate topics as factor analysis, principal components, correlation, cluster analysis, regression, and less commonly used techniques such as discriminant analysis, multivariate analysis of variance (MANOVA), canonical correlation, canonical variate analysis, multivariate analysis of covariance (MANCOVA), multivariate data plotting, and profile analysis. This group by no means exhausts the vast number of topics that could be included in any similiar book. Other related topics not covered in these papers are summarized briefly at the end of this book. It is important to point out to the reader that topics included here do represent a broad segment in the range of current interest.

1.7
Related Texts and Publications

Multivariate quantitative procedures have been applied in geohydrology for some time, but their use has been wrought with some apprehension. Also, many books have presented the theory and background of the methods, and the intent is not to repeat such an exercise. Pertinent references of wide acclaim include undergraduate and graduate school texts such as Draper and Smith (1981), Cooley and Lohnes (1971), Davis (1973, 1986), Mardia et al. (1979), Chatfield and Collins (1980), Johnson and Wichern (1988), Afifi and Clark (1990). Some of the above texts also provide a brief history of methods. There is a notable absence of literature related directly to analysis in geohydrology where large databases abound and continue to grow. What has happened in the last few years to bring about the recent upsurge in the use of multivariate quantitative procedures is that personal computers and workstations have become computationally faster and programming procedures have become more available to the casual PC-user. This change has encouraged large scale application of techniques toward more general geohydrologic problems. Also, recent publications better explain the theory and application of methods to real-world scientific problems. The reader is referred to the texts discussed in this section for more introductory discussions.

1.8
Processing Data

The convenience of computer programs should not make you rush into sophisticated multivariate techniques without examining your data carefully. Missing values should be handled carefully because some programs take blank values to be zeros, whereas others take them to be missing. The analysis of covariance is a valid technique for estimating missing values. Some statistical software programs simply discard any experimental unit which has any of its response variables missing, and no existing software treats missing data adequately. To avoid data entry errors, it is best to enter data by rows of the data matrix, i.e., a line of data for each experimental unit. Large data sets should always be verified by a re-entry of the data, preferably by a different operator. Because the statistical analysis is only as good as the data being analyzed, care and judgement in handling data is required.

A definite trend in making data errors is exhibited by the following order in error pattern. The most common simple errors to look for during data preparation are those concerning outliers, i.e., the extreme values in a data set. The next most common error involves the inversion or interchange of two numbers. The third most common error involves repetition of numbers. The fourth type of error involves having numbers in the wrong column. Computing the range, median and mean of a variable can aid in identifying some of the common errors.

The data for multivariate analysis should always be examined and tested for normality, homogeneity of variances, and multicollinearity. These examinations should be directed at: (1) determination of the suitability of the data for analysis, (2) deciding if transformations are necessary, and (3) deciding what form the data should take. In the context of data analysis, it is important to both prepare the data and examine the data. In some instances, new techniques in simple statistical analysis may be used. These exploratory data techniques include boxplots, scattergrams, histograms, stem-and-leaf plots, and probability plots, to name but a few. The test for normality may be a normal probability plot on variables, tests of skewness and kurtosis, chi-square goodness of fit tests, and/or histograms.

The homogeneity of variances is usually also evaluated using Bartlett's Test of Homogeneity of Variances or another similiar measure. Transformations of data are assumed appropriate by using the Ladder of Powers, but often the most common transformation that is used is the logarithm. Transformations are used to make data more linear, more symmetric, or to achieve constant variance.

1.9
Summary

It is likely that the use of this reference text will cause an increase in the number of applications in the geosciences as computers get faster and procedures are better defined and refined. The proliferation of large databases in geohydrology in the future will require the geoscientist to place greater emphasis on training in these mathematical procedures and statistics in order to extract the most reasonable and meaningful conclusions from the data. It is in this spirit of education that this text is written for the general practitioner. Much impetus for writing this text stems from the knowledge of the author that geoscience is international in scope and similar studies are done worldwide. Because I have discovered that so many of the older text and papers are not available to scientists in other countries and that library and computer resources are scarce, this text is important. This text will help to fulfill scientific needs in the United States and other countries. The synergism of various methods are shown throughout the text. This chapter has briefly introduced the scope of methods that are important in the study of multivariate statistics.

1.10
Supplemental Reading

Davis JC (1986) Statistics and data analysis in geology, 2nd edn. John Wiley, New York
Griffiths JC (1967) Scientific methods in the analysis of sediments. McGraw-Hill, New York

CHAPTER 2

Introduction to Multivariate Statistical Procedures

2.1 Concept

The type of data to be studied is a deciding factor in all statistical methods and is very important in studies using multivariate statistics. Data may be classified as continuous or discrete, normal or non-normal, and based on scales of measurement such as ordinal, nominal, or other.

2.2 Definitions

Normal Distribution. The most important continuous random variable distribution. The graph of frequencies in a class versus probability gives rise to a bell-shaped curve, called the normal curve. A graph of number of frequencies versus class is called a histogram.

Central Limit Theorem. The theorem that states that the average of a large number of independent, identically distributed random variables will have a normal distribution.

Chi-Square Distribution. A mathematical density function that describes the distribution of a chi-square random variable and varies according to degrees of freedom (1 to n). The shape is defined by the degrees of freedom and as n gets large the chi-square distribution approaches the shape of the normal distribution. The chi-square random variable is very important in statistical estimation and forms the basis of chi-square significance tests, usually called chi-square tests.

t-Test. A test based on the t-distribution with m degrees of freedom; the t-distribution approaches the shape of the normal distribution at large degrees of freedom, m. It is called a test of means and is related to the chi-square random variable.

F-Test. A test of variances that is also based on the chi-square random variable and associated distribution. It has m and n degrees of freedom (numerator and denominator degrees of freedom, respectively).

2.3
Purpose and Rationale

As scientists increasingly investigate the world's resources of soil, water, and air, the collected data assume a voluminous space. The databases generated from the many multidisciplinary studies test the limit of the mind to comprehend differences between often hundreds of variables. The best and often only tenuous solution to analyzing such a mass of information is to apply appropriate multivariate quantitative procedures (Table 2.1).

Topics in this text have thus been selected to show the widest application of multivariate quantitative procedures in geology, hydrology, and other environmental sciences. Some discussions in research papers have had a great impact on the prevalent use of certain methods in the geosciences and other discussions are representative of the substantial knowledge gained from other related disciplines. The papers discussed in this text do not represent the first papers on the subject but are those that seem to show the best application and use of selected procedures in geohydrology.

The papers are grouped under seven headings. The headings are based on (1) whether the methods assume a normal distribution for measured data and (2) whether the procedures are variable-directed or individual-directed techniques (called R-mode and Q-mode techniques, respectively). I believe that these discussions will enhance the use of these procedures in future studies. The procedures covered in Part IV do not represent entirely new techniques but are additional applications in the sciences. Other papers that are deemed of merit and published by other scientists may not be covered in this book, and for that reason I will attempt a fuller discussion of related topics when necessary.

The papers discussed in this book come from a wide range of references and journals and the apparent American flavor of the papers is not intentional.

Table 2.1. General classification of multivariate methods

Data type required	Method/analysis type
Normal/non-normal	Correlation (variable)
Non-normal	Principal components (variable)
Normal	Factor analysis (variable)
Non-normal	Cluster analysis (variable/individual)
Normal	Discriminant analysis (individual)
Normal	Tests on covariance matrices (variable)
Normal	Hotelling T^2 (variable)
Normal	Time series analysis (variable)
Normal	Multivariate analysis of variance (variable)
Normal	Canonical correlations (variable)
Normal	Repeated measures experiments (variable)
Non-normal	Multidimensional scaling (individual)
Normal	Multiple regression (variable)
Non-normal	Logistic regression (individual)

Some of the lack of inclusion of very early papers on these subjects in the geosciences can be associated with the lack of computer resources so necessary for computer analysis of large datasets residing in various institutions throughout the world.

2.4
Development of Multivariate Quantitative Procedures

Some of the earliest publications on multivariate quantitative procedures in the sciences were not written for or by geoscientists or hydrologists. Because this work describes research techniques, the material in this book is meant to provide the framework for a number of graduate courses or seminars at universities in this country and internationally. A treatment of matrix algebra (see Appendix), upon which multivariate quantitative procedures are mathematically based is not given in great detail in this book but ample references abound. Examples of earth science data processing are comprehensively described in several books such as Davis (1973, 1986) and Griffiths (1967).

Other text with a more universal application of multivariate quantitative procedures are also valuable such as Draper and Smith (1981); Cooley and Lohnes (1971); Afifi and Azen (1972); Mardia et al. (1979); Chatfield and Collins (1980); Johnson and Wiechern (1988); (Afifi and Clark (1990). These references are especially helpful in providing details on the use of procedures with examples.

2.5
Selection of Methods

In most investigations, the scale of the entity or size of the population makes it impractical to incorporate all of it. If we are to characterize a population, samples must be collected. Special care must be taken in securing the proper experimental design since this affects sampling. The extrapolation of information from samples to populations is the goal of all or most scientific investigations in geology or hydrology.

2.6
Scales of Measurement

In all experimental work the analysis is performed on samples, whether in biology, education, geology, hydrology or other sciences, with the assumption that the sample is representative of the whole unit or population. One of the criteria in choosing a particular statistical procedure is whether data achieves a certain scale (Stevens 1946; Griffiths 1967). The scales of measurement are nominal, ordinal, interval, and ratio. Nominal scale data can be described using counts, mode, contingency, correlation, and independence and include data such as counts of frequency of occurrences of rock types, and other attributes such as color or surface texture. Ordinal data of higher rank than nominal or count data is described by all the previous descriptors plus median, rank, percentiles, and correlations. Ordinal scale data, for example, include Mohs hardness, abundance of heavy minerals, or any other ranked data. The next higher data scale is interval scale and is described by all previous descriptors plus mean, standard deviation, rank order, and correlation. Some examples of this type of data include temperature, sphericity, roundness, oxygen isotope, and absolute time. The highest data scale is ratio and is described by all previous descriptors, plus coefficient of variation and geometric mean. Ratio data has an absolute zero value and is represented by data such as millimeter scale, grams, milliliters, and Kelvin temperature.

2.7 Testing for Normality

2.7.1
Coefficient of Skewness

The coefficient of skewness is often calculated to determine if the distribution is symmetrical or whether it tails to the left (negative) or right (positive). In a general sense, one can look at departures from symmetry of a distribution using the skewness as a measure of normality. One can test by using Pearson's first coefficient of skewness (FCS)(Spiegel 1961):

$$FCS = mean - mode/standard\ deviation. \tag{2.1}$$

One can also test by using Pearson's second coefficient of skewness (SCS):

$$SCS = 3(mean - median)/standard\ deviation \tag{2.2}$$

as a test of normality.
For normal distributions, the moment coefficient of skewness (MCS) is:

$$MCS = a^3 = m_3/S^3 = m_3/(SQRT\ m_2^3) = 0, \tag{2.3}$$

where m_3 = third moment; s = standard deviation; m_1 = mean; m_2 = variance; and m_3 = sum of deviations divided by N.

2.7.2
Coefficient of Kurtosis

The coefficient of kurtosis, CK, is a measure of flatness and may also be tested. For a normal distribution, the CK has a value of approximately 0.263 (Spiegel 1961).

If the data is chosen for transformation, it is important to check that the variable is normal or nearly normal after transformation. This involves finding the transformation that produces skewness and kurtosis values nearest zero, or the transformation showing the fewest outliers. The modules for transformations are available in the popular statistical packages (SAS, SPSSX, BMDP).

2.7.3
Significance Tests – Normality

Graphical displays of data such as histograms and probability plots can be very important in recognizing how data are distributed and if gross errors in the data exist.

New methodologies for applying normality tests now include using probability plots or goodness of fit tests such as chi-square tests.

The standard errors for both kurtosis and skewness can be approximated and then used in a z-test against zero. Even though the equation for kurtosis gives a value of 3 when the distribution is normal, most statistical packages subtract 3 prior to output. The forms of the equations are:

$$z = (S_K - 0)/ss \text{ and } z = (K_u - 0)/sk; \text{ with } S = \text{skewness and } K_u = \text{kurtosis,}$$

where ss and sk are standard errors in the form of:

$$ss = \text{square root of } (6/N) \text{ and } sk = \text{square root of } (24/N).$$

The statistics are very sensitive to sample size and the null hypothesis may be rejected when in fact it should not be. One may generalized that these measures are less important when sample size is large anyway. These measures of normality are output in SPSSx as options in FREQUENCIES and is output as skewness and kurtosis in SAS MEANS and UNIVARIATE. Histograms may also be obtained for large sample size populations to examine distribution characteristics.

2.8
Normal Distribution

The normal or Gaussian distribution is the most common theoretical probability distribution used in statistics. The normal distribution is completely defined by its mean and variance and is graphically defined as having a bell shape. Sixty-eight percent of the values forming the normal distribution lie within one standard deviation of the mean, and 95 percent lie within two standard deviations of the mean. The probability density function for the normal distribution can be found in all standard statistical texts and thus will not be presented here. The log normal distribution also finds extensive use in geohydrology and related sciences, because often a skewed distribution can be described by a lognormal distribution in which case the logarithms of the values are normally distributed.

2.9
Central Limit Theorem

The central limit theorem is very useful and important because it shows that no matter what distribution a group of independent random samples are from, the sample mean of these variables is normally distributed. This is a valid theorem in multivariate analysis as well. The central limit theorem states that the sample mean $(\bar{x})$ is approximately normally distributed with mean $= \mu$ and variance $= \sigma^2/n$, and that the approximation improves with increased sample size. The theorem is very general and makes no assumption about the types of distributions from which the x_i's are derived. How close the approximation is to the normal distribution is sometimes hard to determine, but in most cases, the approximation to the normal distribution is very good for $n > 30$.

The central limit theorem is frequently used to estimate the precision of the sample mean (standard error of the mean) and the confidence in an estimate by allowing error bounds to be determined. For 95 % probability limits, the true population mean is within the specified limits 95 out of 100 times the procedure is applied,(i. e., $\bar{x}$ plus or minus $2S^2/SQRT$ n are the 95 % confidence limits of the mean, where S^2 = Variance).

Data fitting the assumptions of the normal distribution are classed as normally distributed or parametric data and hence may be universally analyzed using parametric statistical techniques. Data that do not fit the general requirements or assumptions of the normal distribution are called nonparametric data and are appropriately only analyzed by nonparametric statistical techniques unless the assumptions of the central limit theorem are invoked, or the data are transformed, as is often done in statistics. Hence one of the criteria for choosing a particular multivariate technique for analysis is whether the data is parametric or nonparametric. This general classification criterion is used in this study, in conjunction with assignment of the data to a scale, or in lieu of some other ground rule of reasoning for relaxing the rules of normality.

2.10
Significance Tests

2.10.1
Chi-Square Goodness of Fit and Probability Plots

The chi-square goodness of fit test, normal probability plot, or other test can be used to test for normality of the data. Examples of these tests are shown later in the text. The chi-square statistic is based on the differences between the actual (observed) frequencies and the frequencies that would be expected to occur if the null hypothesis were true. Data that is fit by the normal probability plot will approach a straight line if the data is normal. The goodness of fit test using the chi-square statistic is used also to test whether a particular distribution is appropriate for a given set of observations.

2.10.2
Significance Tests – Homogeneity of Variances

In most instances, equality of variances is assumed before doing multivariate tests. It is usually then a prerequisite to do one of the procedures before applying multivariate tests. The procedures often used are: (1) Hartley's F-max test, (2) Bartlett's test, (3) Box's test, or (4) Levene's test. Hartley's test requires equal sample size of populations, but in some cases the equal size requirement is relaxed and the calculated statistic is compared with a table of values, as given in Milliken and Johnson (1984). Bartlett's test does not require equal sample sizes. Both Hartley's F-max test and Bartlett's test are quite sensitive to departures from normality and unequal variances. Box's test and Levene's test are much more robust in that they are less sensitive to departures from normality, but are still sensitive to heterogeneous (heteroscedastic) variances (Milliken and Johnson 1984). A one-way analysis of variance is done on the logs of variances for

Box's test, whereas a one-way analysis of variance is done on the variables for Levene's test. Applications of these test are given in Milliken and Johnson (1984) and will not be given here or treated further.

2.11
Division of Procedures – Variable-Directed or Individual-Directed Tests

The second general classification scheme for multivariate quantitative techniques is whether the data are analyzed by variable-directed techniques (R-mode) or techniques that describe relationships between individuals (or samples) called individual-directed techniques (Q-mode). The techniques in this volume have been classified by referring to these ground rules and definitions for data classification and purpose of analysis, i.e., whether for individual or variable comparisons.

2.12
Sampling Design in Multivariate Procedures

The design of sampling programs is an important part of any scientific evaluation or experiment. The procedure for selecting the sample is called the sample survey design. Steps involved in a sampling program cover several aspects, among them are: (1) statement of the objectives, (2) survey design and hypothesis, (3) sample design, (4) data reliability and quality control, and (5) expected results and expense. An ill-conceived hypothesis cannot be studied and leads the investigater to an incorrect conclusion.

In geoscientific analysis, Griffiths (1955) stated that, for comparisons among stratified (subpopulations of) populations, random samples collected in proportion to the magnitudes of the variability encountered in different strata were the most efficient. Curray and Griffiths (1955) concluded that the greatest variability occurred at different sampling levels for different rock types and that the sampling should be proportional to the magnitude of the variability at each stage. Cochran (1960) concluded that well-bedded deposits (stratified) are best sampled by sedimentational unit (strata) samples, unbedded deposits (lack of structure) by channel or grid samples, and poorly-bedded deposits by grid samples. The decision concerning which sampling plan is best suited to a particular outcrop or sampling site should be based upon the sample statistics that are most consistent with population parameters and the type of sampling which achieves consistency with the least number of samples, i.e., most efficiently (Cochran, 1960). For the above reason, it is important to achieve an experimental design that is correct and gives the most unbiased data for the analyses. Ample study of the experimental design is thus very important in all statistical analyses with large data requirements, as is the study and choice of the most efficient sampling procedure.

In an initial study, the correlation analyses can be used to analyse information about which data are more or less redundant, hence providing cost reductions in later phases of the sampling.

2.13
Sample Survey Design

Two popular sampling methods are cluster sampling and stratified sampling. Cluster sampling is a method where the population is divided into clusters; the clusters are randomly chosen and then additional cluster sample members are selected at random to make up the sample. Stratified sampling is a method whereby the population is divided into strata that are as much alike as possible and then the samples are randomly chosen from each strata. These two sampling methods receive wide use and are recommended for most statistical analytic studies.

2.14
Standardization

Standardization is a transformation of a collection of data to standardized, or unitless form, by substracting from each observation the mean of the data set and dividing by the standard deviation. The new variable will then have a mean of zero and a variance of one. This takes care of the problem of comparing two variables measured in different units. The correlation matrix is operated on in algebraic methodology, because it can be regarded as a covariance matrix of standardized variables. The correlation matrix weights all the variables equally and is the desirable starting point in most multivariate studies.

2.15
The Statistical Model

The planning methodology of a multivariate statistical investigation must aim for observations to result from a random sampling process. The following requirements must be satisfied:

1. sampling is random;
2. sampling does not depend on the properties of the sampled entities; and
3. sampling does not represent the personal bias of the investigator.

There are many methods for ensuring randomness in the sampling of data. These randomization procedures fall into a branch of applied statistics known as statistical control. In many cases, especially when observations are received over a longer period of time or in unreasonably large volumes, it is advantageous to divide the population into a number of homogeneous subpopulations and draw random samples from these instead of using a single mass sample from the total population.

2.16
Random Variables

In classical statistics many of the most useful tools are based on the assumption of independent random variables. In simple terms, a random variable denotes a numerical quantity defined in terms of the outcome of an experiment, such as, for example, coin tossing, where the outcome of a coin toss is a random variable.

2.17
Description of Multivariate Quantitative Procedures

In this introduction, I will state briefly the peculiarities of the techniques to be treated in multivariate analysis. I adopt a simple approach to explain only the most important concepts and options available in standard packaged programs. Interested readers should refer to the References for texts that give more comprehensive theoretical treatments of some topics in multivariate analysis. In particular, texts by Chatfield and Collins (1980), Johnson and Wichern (1988), Tabachnick and Fidell (1989) and Affifi and Clark (1990) provide excellent background theory.

There are many general categories of multivariate quantitative techniques or methods (Table 2.1). Multivariate methods allow us to consider changes in several properties of rock, water, or other material simultaneously. The complexity of problems in geohydrology and related sciences often leads one to confront data that are not univariate but rather are multivariate or multidimensional.

The multidimensional space of multivariate data stems from casting variables as dimensions, hence this leads from a concept of "p-variables" to "p-dimensional space", or more generally, many or "n-dimensional" space. With this concept in mind, a short review of multivariate methods is presented in this section. Throughout this text, however, an attempt is made to communicate through example the nature of multivariate data, its distributional assumptions, descriptive parameters, and how to collect, prepare, and analyze data. This abbreviated review is provided to complement the discussions of techniques that are found in other references. The reader is reminded in this discussion that the understanding of simple probability and statistical concepts that were discussed earlier in this text will enhance comprehension, and that an elementary knowledge of statistics is assumed from this point forward.

Once the preliminary examinations of the distribution of the variables are done for depicting outliers, nonnormality, and inequality of variances and no significant violations have been found, then the appropriate hypothesis testing can begin.

2.18
General Significance Testing for Multivariate Data

Multivariate generalizations of univariate t-tests and F-tests are the general procedures used in testing and data analyses.

Most univariate tests are based on the determinant of the $[HE]^{-1}$ matrix, where H is the hypothesis sums-of-squares and cross-products matrix (SPSS, 1985). The determinant, denoted by $||$, is a measure of the generalized variance, or dispersion, of a matrix. This determinant can be calculated as the product of eigenvalues of a matrix, since each eigenvalue λ represents a portion of the generalized variance. From a different perspective, the process of extracting eigenvalues can be viewed as a principal components analysis of the $[HE]^{-1}$ matrix. There are a number of test statistics for evaluating multivariate differences based on the eigenvalues of the $[HE]^{-1}$ matrix. The multivariate analysis of variance procedures in, for example, Statistical Package for the Social Sciences (SPSS; Norusis 1985) uses various test criteria:

1. Pillai's trace, $\qquad V = \sum_{i=1}^{s} \dfrac{1}{1+\lambda_i}\,;$

2. Wilks' λ, $\qquad W = \prod_{i=1}^{s} \dfrac{1}{1+\lambda_i}\,;$

3. Hotelling's trace, $\qquad T = \sum_{i=1}^{s} \lambda_i\,;\quad$ and

4. Roy's largest root, $\qquad R = \dfrac{1}{1+\lambda_{max}}\,;$

where λ_{max} is the largest eigenvalue, λ_i is the ith eigenvalue, and s is the number of nonzero eigenvalues of the $[HE]^{-1}$ matrix. These four test statistics have different distributions, but they can be transformed into statistics that have approximately an F distribution. Tables of the exact distributions of the statistics are also available. A fuller treatment of this subject is not warranted at this time and the reader is referred to Norusis (1985), Johnson and Wichern (1988) or Mardia et al. (1979) for more details.

2.19
Summary

This chapter has introduced and reviewed the background of multivariate methods, the selection process for methods, types of data, data handling procedures, sampling methods, and experimental designs that are the basis of the applied procedures. Comprehension of these concepts is very important in choosing the correct technique for analysis of the data.

2.20
Supplemental Reading

Afifi A, Clark V (1990) Computer-aided multivariate analysis. Van Nostrand Reinhold, New York
Chatfield C, Collins AJ (1980) Introduction to multivariate analysis. Chapman and Hall, New York

Variable-Directed Procedures Based on Normal Distribution Assumptions

CHAPTER 3

Correlation

3.1
Concept

In many studies, there may be many independent and dependent variables that are measured and must be analyzed. The basis of multivariate procedures derives from an n-dimensional random variable, sometimes called an independent random vector composed of univariate random variables in the form of matrices.

Correlation measures the size and direction of the relationship between two variables. The most frequently used measure is the Pearson product-moment correlation coefficient, r. The Pearson r is the covariance between X and Y relative to the square root of the X and Y variances.

3.2
Definitions

Correlation. The degree of association between two quantities. The association is scaled and is always between 1 and –1.

Covariance. The degree of dependence between two quantities, x and y, denoting how they vary (positively or negatively). Values are not constrained between 1 and –1, and covariance represents the numerator in the calculation of the correlation coefficient for variables i and j.

Estimator. Quantity based on observations of a sample whose value is taken as an indicator of the value of an unknown population parameter, i.e., the average. Estimators may be: (1) "consistent" in that they tend to converge toward the true value with increasing sample size n; (2) "maximum likelihood estimators" (MLE) wherein if the true value of the unknown parameter has this value, then the probability of obtaining the sample is maximized (such as sample average is MLE for mean of population and we can also calculate MLEs for variance, binomial probability of success, correlation, and others); (3) "unbiased" estimators wherein the expected value is equal to the true value of the parameter it is trying to estimate. The process of using observations of a sample to estimate the properties of a population is statistical inference.

3.3
Objective of Procedures

A logical approach to analysis of multivariate data is begun with the reduction of the [n × p] data matrix to a [p × p] matrix of correlation coefficients. A basic feature of this correlation is that the values of the variables have been transformed so that each mean is zero ($\bar{x} = 0$) and the variance σ^2, is equal to one. This permits a quick comparison of the different measured variables. If significant correlations in the data are not found during this so called, preanalysis phase, it is not necessary to proceed to principal-components analysis or factor analysis and a new suite of variables with larger correlations must be determined and analyzed. Use of correlation coefficients requires testing of the significance of the statistical estimators before drawing inference.

3.4
Problems

Multicollinearity (very highly correlated variables) and singularity (perfectly correlated variables) must be addressed. This leads to an almost zero-valued determinant and division leads to unstable numbers in the inverted matrix. The check is the squared multiple correlation. Redundant variables, in most statistical analyses except factor analysis or principal components analysis, are to be avoided (sometimes correlations of 0.70 or higher should be analyzed and certain data excluded from further analysis).

3.5
Significance Tests for Correlations

For degrees of freedom, $v = n - 2$, we test the hypothesis, Ho: that the correlation coefficient (r) is not significantly different from zero (no correlation.) Zero-order correlations are then analyzed appropriately for meaning of correlations and it is determined whether factor analysis or principal components analysis is appropriate for further study of this data. The sample size and size of the correlation coefficient are the important parameters to be evaluated. The correlation coefficient can be tested against a table of values such as from Arkin and Colton (1962, p. 155).

When several variables are interrelated, the simple correlation coefficient between pairs of variables may give misleading results. This is a limitation that can be treated using the partial correlation coefficient. The usefulness of this technique is to remove the effect of masking, and this procedure is done by holding all the closely related variables fixed.

3.6
Interpretation

The correlation coefficient measures the strength of the linear relationship between two variables x_i and y_j. Correlations close to zero indicate that no linear relation exists between the two variables, but do not necessarily imply that no relation exists, (for instance a nonlinear relation may exist which is not accounted for), unless the two variables have a joint bivariate normal distribution. When the sample size is very large, virtually all sample correlation coefficients will be significantly different from zero, but the question arises as to the importance of the correlations. This importance is determined by looking at the practical importance, i. e., whether the correlation is large (usually greater than 0.7 as a general rule) or is criterion–based and determined from a table value based on sample size (such as Arkin and Colton 1962). The caution here is that correlations may not be significant or important if the sample size is less than about 12, and small sample size can lead to spurious results that may not be reliable.

3.7
Numerical Example on Cave Lithology

A correlation analysis of variables describing cave lithology was done by Rauch and White (1970). As previously discussed, the correlation is an indication of the degree of association between two quantities, and its value is always between –1 and +1. A related term, the covariance is a measure of dependence and is also an indication of the degree of association between two quantities. It is related to the correlation coefficient, but is not constrained between –1 and +1.

Rauch and White (1970) used correlations to analyze and understand the lithologic controls on the distribution of solution porosity in Middle Ordovician carbonate-rock aquifers in the Valley and Ridge province of central Pennsylvania. The study encompassed areas in the Nittany, Brush, Penns, and Sugar valleys. Because it is generally known that the movement of ground water in karst regions is through solution conduits of considerable size, this paper evaluated correlations between variables such as cave column and cave length, the distribution of cave volume, petrographic and chemical variables, and provided a comprehensive data table that was operated on through a correlation analysis. Extensive cave development was found to exist in limestones and not in dolomites. Table 3.1 shows the correlations between specific speological variables measured on the upper argillaceous section for 118 samples. Similiar patterns in the data were found for measured variables in the middle cavernous section and lower dolomitic section. This paper has received a significant amount of attention as a good reference to the study of carbonate rocks and cave volume due to its multivariate nature.

Based on the analysis, Rauch and White (1970) established that non-dolomitized low magnesium rocks have the most solution cavities in the carbonate rocks studied. Low silica rocks are found to correlate with cave development, whereas the distribution of Al_2O_3 is less significant. Principal components analysis was also done as part of their study, but results are not provided here. The conclusion is that cave development is inhibited by high concentrations of SiO_2, Al_2O_3, dolomite, sparite, and impurities, or by very low dolomite concentrations. Other later studies (Brown, 1977, 1993) support the conclusions found by Rauch and White (1970).

Table 3.1. Correlation coefficients for hydrologic data from caves. (Modified from Rauch and White 1970) Reprinted by permission of Water Resources Research. Copyright 1970. All rights reserved

Upper argillaceous section (118 samples)

Variables	Variables							
	1	2	3	4	5	6	7	8
2	−0.959							
3	−0.131	−0.061						
4	−0.116	0.089	0.036					
5	−0.399	0.223	−0.072	−0.117				
6	−0.008	−0.051	−0.049	−0.084	0.305			
7	−0.095	−0.003	0.136	−0.010	0.321	0.115		
8	−0.022	−0.095	0.191	0.105	0.291	−0.004	0.640	
9	−0.117	0.075	0.012	0.165	0.186	−0.014	−0.093	0.183

Variables (values in percent except Micrite Grain Size)

1 Micrite	6 Micrite Grain Size
2 Sparite	7 Silica
3 Quartz and Feldspar	8 Aluminium Oxide
4 Pyrite	9 Magnesium Oxide
5 Other	

3.8
Numerical Example on Carbonate-Rock Geohydrology

A correlation analysis on variables describing carbonate–rock aquifers was done in this example. Table 3.2 contains the correlations determined during studies on carbonate rock aquifers in central Pennsylvania (Brown 1977, 1993). The rocks studied are the Lower Ordovician Beekmantown carbonates as represented by the Stonehenge limestone, Nittany dolomite, Axemann limestone, and upper and lower Bellefonte dolomite. Variables that were measured include: bulk density, porosity, permeability, insoluble residue, grain size, total carbonate, percent calcite as CaO, percent magnesium as MgO. The data shows that correlations between reservoir and petrographic properties are important in establishing the potential of a carbonate rock aquifer to produce water, or a reservoir rock to produce oil and gas. The data from the two correlation tables that were discussed earlier both indicate that MgO is a significant factor in understanding porosity and permeability of carbonate rocks, because high magnesium limestones and dolomites were found to have the highest porosity and permeability of rocks studied.

3.9
Numerical Example on Fluvial Sediment Geochemistry

A geostatistical analysis of fluvial sediments was undertaken to determine the source area for certain sediments.

Table 3.2. Correlation matrix of variables from lower Beekmantown carbonates in central Pennsylvania. (Modified from Brown 1977, 1993) (Reprinted from Chemical Geology with kind permission from Elsevier Science – NL, Sara Burgerhartstraat 25, 1055 KV Amsterdam, The Netherlands. Copyright 1993)

| Variables | Variables | | | | | | | | | |
	1	2	3	4	5	6	7	8	9	10
2	−0.05									
3	−0.13	0.48								
4	0.07	0.43	−0.15							
5	−0.07	−0.43	0.15	−1.0						
6	0.06	−0.37	−0.60	0.22	−0.22					
7	−0.10	−0.18	0.21	−0.29	0.29	−0.01				
8	0.01	−0.36	−0.61	0.21	−0.21	0.99	−0.02			
9	−0.15	−0.25	0.15	−0.34	0.34	0.04	0.96	0.05		
10	−0.73	−0.42	−0.03	−0.47	0.47	0.16	0.38	0.20	0.44	
11	0.78	0.36	0.02	0.31	−0.32	−0.19	−0.35	−0.23	−0.41	−0.98

Variables

1 Bulk density, g/cc	7 A-axis standard deviation
2 Porosity%	8 B-axis grain size (B-ϕ)
3 Log permeability, millidarcies	9 B-axis standard deviation
4 Insoluble residue%	10 Calcite as calcium oxide%
5 Carbonate%	11 Dolomite as magnesium oxide%
6 A-axis grain size(A-ϕ)	

Many geological studies have included correlations as an important analytical tool to formulate meaningful conclusions when analysing multivariate data. Ali (1984) analyzed 134 samples of sediments from the Euphrates and Tigris Rivers and Pliocene molasse sediments of the Bakhtiari formation. The correlation coefficients were used to systematically provide an association of source area to chemical composition. Such an association is usually found to be very significant in exploration activities for water or minerals. The conclusion is that the relatively high concentrations of Cr, Ni, V, Cu, and Fe and lower concentrations of Ga and Al in the Pliocene Bakhtiari formation and the recent Euphrates-Tigris sediments suggest an origin in the weathered basic igneous rocks from areas of the Torons and Zagros ranges in north and northeast Iraq. Tables 3.3, 3.4, 3.5, and 3.6 show the values of the correlation coefficients used in the study of the fluvial sediments of the Euphrates and Tigris rivers.

Ali (1983) found that the analysed whole samples and the clay-sized fraction of both the Bakhtiari and recent sediments are similiar and are probably derived from similiar source rocks. He found that the source areas of the Euphrates River sediments are the highlands in Turkey, Iran, and the nappe zone of Iraq which are composed of clastic rocks. He also stated that relatively higher concentrations of Cr, Ni, V, Cu, Zr, Fe, and Ti were found to occur within the Euphrates and Tigris sediments than in the Bakhtiari sediments, while the concentrations of Rb, Sr, B, Si, and K are rather similiar in all sediments. A tendency was found for lower concentrations of Ga and Al in the Bakhtiari than in the Euphrates and Tigris samples.

Table 3.3. Matrix for the correlation coefficients of the elements for bulk molasse sediments. (Modified from Ali 1984) (Reprinted from Chemical Geology with kind permission of Elsevier Science – NL, Sara Burgerhartstraat 25, 1055 KV Amsterdam, The Netherlands. Copyright 1984)

Variables	1	2	3	4	5	6	7	8	9	10	11	12	13	14
1 Rb														
2 Sr	0.358													
3 B	0.224	-0.207												
4 Ga	0.449	0.150	0.473											
5 Cr	0.162	0.146	0.148	0.103										
6 Ni	0.214	0.064	0.207	0.093	0.516									
7 V	0.163	0.174	0.512	0.363	0.127	0.301								
8 Cu	0.171	0.058	0.155	0.004	0.065	0.244	0.416							
9 Zr	0.050	0.333	0.062	0.004	0.173	0.124	0.045	0.289						
10 SiO$_2$	0.382	0.146	0.087	0.062	-0.096	-0.178	0.185	0.021	0.034					
11 Al$_2$O$_3$	0.458	0.006	0.180	0.121	-0.025	-0.138	0.461	-0.083	0.254	0.519				
12 TiO$_2$	0.359	0.014	0.103	-0.062	0.065	-0.076	0.230	-0.005	0.223	0.291	0.487			
13 K$_2$O	0.104	-0.297	0.315	0.231	-0.169	-0.094	0.292	0.156	-0.093	0.194	0.452	0.225		
14 Fe$_2$O$_3$	0.122	-0.039	0.142	0.170	0.173	0.948	0.143	-0.099	-0.412	0.357	-0.024	0.094	0.207	

Table 3.4. Correlation matrix for clay-sized molasse sediments (Modified from Ali 1984) (Reprinted from Chemical Geology with kind permission of Elsevier Science – NL, Sara Burgerhartstraat 25, 1055 KV Amsterdam, The Netherlands. Copyright 1984)

Variables	1	2	3	4	5	6	7	8	9	10	11	12	13	14
1 Rb														
2 Sr	0.034													
3 B	0.526	0.063												
4 Ga	0.717	0.111	0.546											
5 Cr	0.002	0.176	0.179	0.112										
6 Ni	0.194	0.092	-0.217	-0.127	0.521									
7 V	0.350	-0.061	0.256	0.633	0.252	0.314								
8 Cu	-0.118	0.051	0.139	-0.109	0.321	0.479	0.021							
9 Zr	0.403	0.500	0.476	0.376	0.183	-0.086	0.161	0.172						
10 SiO$_2$	0.454	0.045	0.457	0.524	0.060	-0.103	0.331	-0.145	0.500					
11 Al$_2$O$_3$	0.321	0.002	0.518	0.360	0.016	-0.177	0.197	0.018	0.204	0.752				
12 TiO$_2$	0.451	0.027	0.531	0.603	0.094	-0.251	0.217	0.004	0.463	0.831	0.796			
13 K$_2$O	0.354	-0.002	0.550	0.526	0.083	-0.092	0.286	0.008	0.302	0.743	0.897	0.796		
14 Fe$_2$O$_3$	0.001	-0.032	0.205	-0.061	0.046	0.042	0.041	0.009	-0.039	0.024	0.201	0.195	0.218	

Table 3.5. Correlation matrix for bulk recent fluvial sediments (Modified from Ali 1984) (Reprinted from Chemical Geology with kind permission of Elsevier Science – NL, Sara Burgerhartstraat 25, 1055 KV Amsterdam, The Netherlands. Copyright 1984)

Variables	1	2	3	4	5	6	7	8	9	10	11	12	13	14
1 Rb														
2 Sr	0.289													
3 B	0.138	-0.131												
4 Ga	0.244	0.368	0.328											
5 Cr	0.166	0.002	-0.059	0.170										
6 Ni	-0.266	-0.217	-0.068	-0.063	-0.045									
7 V	-0.063	0.105	0.201	0.173	0.007	0.290								
8 Cu	0.105	-0.124	-0.001	0.032	-0.325	0.050	0.202							
9 Zr	0.259	0.319	-0.105	0.180	0.290	-0.384	-0.030	-0.096						
10 SiO_2	0.031	-0.109	0.253	0.167	0.305	-0.545	-0.012	-0.118	0.090					
11 Al_2O_3	0.435	0.457	0.066	0.677	0.089	-0.324	0.240	0.182	0.365	0.233				
12 TiO_2	0.288	0.194	0.318	0.477	0.292	-0.160	0.153	0.099	0.148	0.177	0.471			
13 K_2O	-0.099	-0.312	0.009	-0.132	-0.123	0.544	0.051	-0.023	-0.156	-0.151	-0.285	-0.233		
14 Fe_2O_3	-0.135	-0.342	-0.011	-0.350	0.028	0.345	0.314	0.180	-0.094	-0.010	-0.148	-0.155	0.224	

Table 3.6. Correlation matrix for clay-sized recent fluvial sediments. (Modified from Ali 1984) (Reprinted from Chemical Geology with kind permission of Elsevier Science – NL, Sara Burgerhartstraat 25, 1055 KV Amsterdam, The Netherlands. Copyright 1984)

Variables	1	2	3	4	5	6	7	8	9	10	11	12	13	14
1 Rb														
2 Sr	0.014													
3 B	0.182	-0.201												
4 Ga	0.111	-0.118	0.310											
5 Cr	0.321	0.104	-0.370	0.005										
6 Ni	0.187	0.308	-0.255	0.169	0.219									
7 V	0.160	-0.175	0.280	0.540	0.293	0.241								
8 Cu	0.352	-0.248	-0.127	-0.103	-0.096	0.375	0.036							
9 Zr	0.115	0.433	-0.094	0.097	0.305	0.406	-0.133	0.233						
10 SiO_2	0.270	-0.287	0.162	0.329	-0.284	-0.39	-0.58	0.272	-0.045					
11 Al_2O_3	0.406	-0.031	0.043	0.38	0.178	0.306	0.055	0.269	-0.002	0.466				
12 TiO_2	0.076	0.129	0.012	0.422	0.122	0.261	0.170	-0.160	0.144	0.232	0.452			
13 K_2O	0.422	-0.199	-0.348	-0.352	0.422	0.749	0.129	0.484	-0.197	-0.491	0.555	0.442		
14 Fe_2O_3	0.041	-0.317	0.182	0.247	-0.011	0.294	0.441	0.069	-0.170	0.043	0.107	-0.213	0.246	

3.10
Numerical Example – Discussion

An important aspect of correlation analysis is how we formulate the analysis, and the formulation is very important if we are dealing with percentage data. Butler (1974) discussed the significance of a problem inherent in looking at correlations between percentages which are often a large part of geologic data. He concluded that if data are percentages of the same whole, the significance of the observed pattern is difficult to assess, as are the correlations. This example follows.

Table 3.7 illustrates very well the discrepancies in correlations between a closed data set (sum to 100%) and an open data set. The variance-covariance tables and the correlation tables are different for the open and closed data sets.

Table 3.7. Covariance-variance and correlation matrices for open (A,C,E and G) and closed (B, D, F, and H) datasets. (Modified from Butler 1974) Reprinted by Permission of Geological Education. Copyright 1974

(A) Open form	Matrix X Variables				Sum of
Samples	A	B	C	D	rows
1	80	21	44	50	195
2	88	51	53	56	248
3	24	30	8	99	161
4	66	31	22	83	202
Sum of columns	258	133	127	288	
Column means	64.5	33.3	31.8	72.0	
Variance	811.7	160.3	420.3	530	
Standard deviation	28.5	12.7	20.5	23.0	
Coefficient of variation	0.441	0.381	0.646	0.319	
(B) Closed form	Matrix Y Variables				Sum of
Samples	A	B	C	D	rows
1	41.0	10.8	22.6	25.6	100.00
2	35.5	20.6	21.4	22.5	100.00
3	14.9	18.6	5.0	61.5	100.00
4	32.7	15.3	10.9	41.1	100.00
Sum of columns	124.1	65.3	59.9	150.7	
Column means	31.0	16.3	15.0	37.7	
Variance	128.5	18.4	71.6	320.0	
Standard deviation	11.3	4.3	8.5	17.9	
Coefficient of variation	0.365	0.264	0.567	0.475	

Table 3.7 (continued)

| (C) Open form | Deviation Matrix X Variables | | | | Sum of rows |
	A	B	C	D	
Samples					
1	15.5	−12.25	12.25	−22.0	−6.5
2	23.5	17.75	21.25	−16.0	−46.5
3	−40.5	−3.25	−23.75	27.0	−40.5
4	1.5	−2.25	−9.75	11.0	0.5
Column sum	0.0	0.0	0.0	0.0	−93.0

| (D) Closed form | Deviation Matrix Y Variables | | | | Sum of rows |
	A	B	C	D	
Samples					
1	10.0	−5.5	7.6	−12.1	0.0
2	4.5	4.3	6.4	15.2	0.0
3	−16.2	2.3	−10.0	23.9	0.0
4	1.7	−1.1	−4.0	3.4	0.0
Column sum	0.0	0.0	0.0	0.0	0.0

| (E) Open form | Variance–covariance Matrix X Variables | | | | Sum of rows |
	A	B	C	D	
A	811.7	118.5	545.5	−598.0	0.0
B	118.5	160.2	108.8	42.3	345.2
C	545.5	108.8	420.3	−452.7	621.8
D	−598.0	−42.3	−452.7	530.0	536.0

| (F) Closed form | Variance-Covariance Matrix Y Variables | | | | Sum of rows |
	A	B	C	D	
A	128.5	−24.9	86.7	−190.3	0.0
B	−24.9	18.4	−10.9	17.5	0.0
C	86.7	−10.9	71.5	−147.3	0.0
D	−190.3	17.5	−147.3	320.0	0.0

| (G) Open form | Correlation coefficient Matrix X Variables | | | |
	A	B	C	D
A	1.00	0.33	0.93	−0.91
B	0.33	1.00	0.42	−0.15
C	0.93	0.12	1.00	−0.96
D	−0.91	−0.15	−0.96	1.00

| (H) Closed form | Correlation coefficient Matrix Y Variables | | | |
	A	B	C	D
A	−1.00	−0.49	0.90	−0.94
B	−0.49	1.00	−0.29	−0.23
C	0.90	−0.29	1.00	−0.97
D	−0.94	−0.23	−0.97	1.00

3.11
Summary

Preliminary analyses of variable relationships are best initiated through a correlation analysis of data. This step gives an indication of the need to run other analyses such as factor analysis or other multivariate procedures on the dataset.

3.12
Supplemental Reading

Griffiths JC (1967) Scientific methods in analysis of sediments. McGraw-Hill, New York

CHAPTER 4

Factor Analysis

4.1
Concept

Factor analysis is used to form a subset of uncorrelated theoretical variables called factors that adequately explain the variation in the original variable set.

4.2
Matrices

The matrices of interest in factor analysis are: (1) matrices of correlations (between variables, between factors, between variables and factors); (2) matrices of standard scores (on variables, and on factors); (3) matrices of regression weights (for producing scores on factors from scores on variables); (4) pattern matrix (of unique relationships between factors and variables after oblique rotation; and (5) matrix of partial correlations (since bivariate correlations may be suspect, pairwise correlations are adjusted for effects of all other variables).

4.3
Definitions

Correlation. A $p \times p$ matrix of variable correlations.

Variable Matrix. An N sample $\times$ p matrix of standardized observed variable scores.

Factor-Score Matrix. An $N \times m$ factors matrix of standard scores on factors.

Factor Loading Matrix. A $p \times m$ matrix of correlations between variables and factors.

Factor-Score Coefficient Matrix. A $p \times m$ matrix of regression weights used to generate factor scores.

Structure Matrix. A $p \times m$ matrix of correlations between variables and (correlated or nonorthogonal) factors.

Factor Correlation Matrix. A $m \times m$ matrix of correlations among factors.

Eigenvalues. The lengths of the axes of an n-dimensional ellipsoid; the sum of the eigenvalues equal the trace of variance-covariance matrix which is equal to total variance of data set.

Eigenvectors of Matrix. The principal axes of an n-dimensional ellipsoid (containing defining points).

Communalities. Sum of squared factor loadings for row.

Factor Scores. Product of standardized scores on variables and factor score co-efficients.

Reproduced Correlation Matrix. $[\mathrm{Rcm}_1] = [\mathrm{R}] \times [\mathrm{Rtranspose}] = [\mathrm{R}]^2$.

Residual Correlation Matrix. $[\mathrm{Rcm}_2] = [\mathrm{Robs}] - [\mathrm{R}]^2$.

Factor Correlation Matrix. Correlation of first and second factors after oblique rotation.

Predicted Standardized Scores on Variables. Product of scores on factors weighted by factor loadings.

Unrotated Factor Loadings. Original factor loadings from variables.

Orthogonal Rotation. Rotation as such to keep factors uncorrelated (most popular methods are varimax, quartimax, and equamax).

Principal Components. The eigenvectors of a variance-covariance matrix.

4.4
Contrasts Between Factor and Principal Components Analyses

Factor analysis (denoted FA) produces what may be called factors and principal components analysis (PCA), which is a similar method, produces principal components, which are eigenvectors of a variance-covariance matrix. This procedure may be looked at, for example, as a redefinition of the correlation matrix, i.e., mathematically it is a process of matrix substitution. The FA and PCA procedures are similar except for preparation of the observed correlation matrix for extraction. The difference between FA and PCA is in the variance that is analyzed. In PCA, all the variance in the observed variables is analyzed, but in FA only the shared variance of variables (communalities) is analyzed and attempts are made to estimate and eliminate variance due to error and variance that is unique to each variable.

Mathematically, the difference between PCA and FA involves the contents of the positive diagonal in the correlation matrix (the diagonal that contains the correlation between a variable and itself). In either PCA or FA, the variance that is analyzed is the sum of the values in the positive diagonal. In PCA, ones are in the diagonal and there is as much variance to be analyzed as there are variables; each variable contributes a unit of variance to the positive diagonal of the correlation matrix. All the variance is distributed to the components, including error and unique variance for each observed variable. So if all components are retained, PCA duplicates exactly the observed correlation matrix and the standard scores of the observed variables. In FA, only the variance that each variable shares with other observed variables is available for analysis. Exclusion of error and unique variance from FA is based on the belief that such variance only confuses the picture of underlying processes. Shared variance is estimated by communalities, values between zero and one that are inserted into the positive diagonal of the correlation matrix. The solution of FA concentrates on variables with high communality values (Tabachnick and Fidell 1989).

If examination of data leads to conclusions that significant correlations are found to exist in the correlation data table, it is appropriate and feasible to proceed to more advanced procedures such as factor analysis. If significant correlations are not found in the data, the analysis should provide for the collection of more samples, and the act of continuing with advanced analysis is not feasible, nor advised.

Factor analysis is also similar to principal components analysis in other ways also in that it is a technique for examining the interrelationships among a set of variables. The two techniques are different and should not be confused. Factor analysis is more concerned with explaining the covariance structure of the variables, whereas principal components analysis is more concerned with explaining the variability in the variables. Both of these techniques differ from regression analysis in that we do not have a dependent variable to be explained by a set of independent variables. However, principal components analysis and factor analysis also differ from each other. In principal components analysis, the major objective is to select a number of components that explain as much of the total variance as possible. Speaking from a mathematical standpoint, the principal components are the eigenvectors of a variance-covariance matrix and are usually the starting point of all factor analysis models. For example, in a two-dimensional case, the variance-covariance structure of the data can be expressed as a 2×2 matrix, or as two vectors. The values of the principal components for a given individual are relatively simple to compute and interpret. On the other hand, the factors obtained in factor analysis are selected mainly to explain the interrelationships among the original variables. Ideally, the number of factors expected is known in advance. The major emphasis is placed on obtaining easily understandable factors that convey the essential information contained in the original set of variables.

Areas of application of factor analysis are similar to those mentioned for principal components analysis. Original applications have come mainly from other sciences, particularly education, psychology, and biology. The technique has been used chiefly to explore the underlying structure of a set of variables. It has also been used in assessing what items to include in studies and to explore the interrelationships among the scores on different items. A certain degree of resistance to using factor analysis in some disciplines has been prevalent, perhaps because of the heuristic nature of the technique and the special jargon employed. The early applications of the technique involved an analysis of abstract concepts, but currently in science, analyses involve the description of physical concepts and processes. Also, the large number of methods available to perform factor analysis leaves some investigators uncertain of their results.

4.5
Significance Tests – Factor Analysis Decision Tests

A more sophisticated test of the adequacy of doing other tests such as factor analysis is given by the Kaiser-Meyer-Olkin (KMO) measure of sampling adequacy (Norusis 1985). The KMO is an index for comparing the magnitudes of the observed correlation coefficients to the magnitudes of the partial correlation coefficients. Small values of the KMO measure indicate that a technique such as factor analysis may not be a good idea. Kaiser (1974) has indicated that KMOs below 0.5 are not acceptable.

4.5.1
Significance Tests – Bartlett's Test of Sphericity

Bartlett's test of sphericity can be used to test whether the correlation matrix is an identity, i. e., that all diagonal terms are one and all off-diagonal terms are zero (Norusis, 1985). If the value of the test statistic for sphericity (based on a chi-square transformation of the determinant of the correlation matrix) is large and the associated significance is small, it is unlikely that the correlation matrix is an identity, and factor analysis would not be recommended as a study method in this case. Multivariate normality is an assumption for the data for this test (see Sect. 2.7).

4.5.2
Generalizations About Correlations and Factor Analysis

The interrelations or correlations among variables can be depicted graphically. If one were to look at a graph of a problem depicting the relation between variables, the depiction is somewhat oversimplified. If one plots the variables on a circular diagram, a vector (arrow) is drawn representing each variable (Fig. 4.1) and the angle between any two variables as shown reflects the correlation between them. For analogy, if we take the cosine of the angle between them as their correlation (identical coefficients if standardized variables used, otherwise very similiar coefficients if unstandardized), the smaller the angle between any two vectors, the greater is their correlation. In one sense the cosine is a measure of the 'great circle' distance between samples which lie on a hypersphere. For example, if the angle between any two is ten degrees, than its correlation is equal to 0.985, which is the cosine of ten degrees (Table 4.1). For those variables having angles of 80 degrees between them, the correlation is 0.174, which is the cosine of 80 degrees. Measuring the angles for all pairs of variables would then give a table of correlations. If one next decides to look at Q-mode factor analysis, which often uses the cosine method, one can see from Fig. 4.1 how the variables can correlate with the factors in the same way as variables with one another. Table 4.1 gives the correlations and factor loadings for a hypothetical dataset.

Figure 4.1 also shows what happens when we rotate matrices during factor analysis, a procedure to be discussed later. We tend to increase or decrease the

Figure 4.1. Graphic portrayal of A intercorrelations, B unrotated matrix, and C rotated matrix for data in Table 4.1. The correlation between any two vectors is equal to the cosine of the angle between them. Variables are vectors A,..., E, and factors are vectors I and II. (Williams 1979)

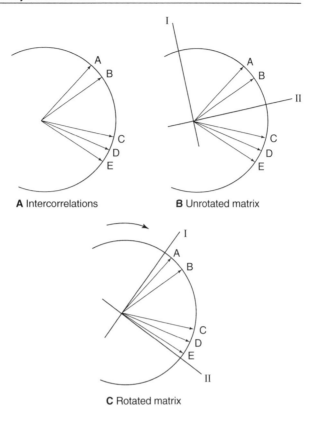

A Intercorrelations **B** Unrotated matrix

C Rotated matrix

angles developed between any two variables and between any factor and variable. The cosine is found as in the earlier discussion, but now in terms of the factors as reference points, the correlations are called factor loadings. The factors are defined as maintaining an angle of 90° between them, i.e., they are uncorrelated because the angle of separation is 90° and the cosine of 90° is zero.

If we look in more depth into the correlation of two variables on two factors, for example, in factor analysis, there is a relationship found that can be clearly defined. The correlation between variables A and B, designated r(AB), is equal to the sum of the products of the two variable loadings on each factor (for either the rotated or unrotated matrix) according to Eq. (4.1):

$$r(AB) = [A_1B_1 + A_2B_2] \tag{4.1}$$

where A_1 means the loading of variable A on factor 1, and B_2 denotes the loading of variable B on Factor 2, and so on. For example, if the correlation between A and B is 0.985, and we look at factor 1 and factor 2 loadings, we would calculate a correlation as follows:

$$0.985 = (0.996 \times 0.966) + (0.087 \times 0.259)$$

and this would hold for both the rotated or unrotated matrix loadings.

Table 4.1. Example of (A) intercorrelation matrix, (B) unrotated factor matrix, and (C) rotated factor matrix (Williams 1979)

(A) Intercorrelations

		Variables			
		B	C	D	E
	A	0.985	0.500	0.342	0.174
Variables:	B		0.642	0.500	0.342
	C			0.985	0.940
	D				0.985

(B) Unrotated factor matrix

		Factors	
		1	2
	A	0.574	0.819
Variables:	B	0.423	0.906
	C	−0.423	0.906
	D	−0.574	0.819
	E	−0.707	0.707

(C) Rotated factor matrix

		Factors	
		1	2
	A	0.996	0.087
	B	0.966	0.259
	C	0.423	0.906
Variables:	D	0.259	0.966
	E	0.087	0.996

4.6
Determining Significance of Output

The most important parameters to be evaluated in factor analysis are communality, factor loadings, size of eigenvalues, and sum of explained variance. Communality is the portion of variance explained by common factors, i.e, the sum of λ_{jk^2}. The factor loadings are simply the weights loaded on the factors. The sum of the factor loadings squared is equal to the eigenvalue which is the variance explained by a factor. If there are five variables, it is expected that the variance should equal five. This is often not the case because there error variance and nonerror variance (specific to variables) are not included in the factor structure. Putting ones on the diagonal of the correlation matrix is standard procedure to indicate that a variable is totally correlated with itself, but if we know the measurement error variance from prior research, we can use some value less than one.

The factor loading, for example, λ_{jk}, is the loading of the jth response variable on the kth factor. Eigenvalues are taken to be important if values are larger than or equal to one. The sum of explained variation should total at least 90 % and this level of significance denotes when factors are probably not important and should be eliminated before rotation of data axes is done. Rotation is often done because the matrix of factor loadings is not unique or easily explained. Multicollinearity can be a problem in factor analysis where singularity (perfect correlation of some variables) exists, but in PCA, multicollinearity is not a problem because there is not a need to invert a matrix.

In FA, the correlation matrix produced by the observed variables is called the 'observed correlation matrix'. The correlation matrix produced from factors is called the 'reproduced correlation matrix'. The difference between the observed and reproduced correlation matrices is the 'residual correlation matrix' (Tabachnick and Fidell 1989). In this reference frame, the factor-analytic solution is a redefinition of the correlation matrix in terms of a factor matrix.

4.6.1
Significance Tests – Factor Loadings

The factor loadings can be tested in the same manner as the correlation coefficient, i. e., by using a table of test values such as from Arkin and Colton (1962). The criterion is based on degrees of freedom, n-2, and is used in the same manner as a test of the correlation coefficient.

4.6.2
Significance Tests – Amount of Explained Variation

A factor is taken as being significant if it adds a significant amount of variation or its eigenvalue is greater than one. A scree plot of factors versus explained variation may also be used and the point at which the derived curve flattens is a point of no return, i. e., factors beyond this point are unimportant.

4.7
Extraction Techniques for Factors

The principal extraction techniques of popular statistical packages include: (1) principal components, which mathematically determines an empirical solution with common, unique, and error variance mixed into components; (2) principal factors, which estimates communalities in an attempt to eliminate unique and error variance from factors; (3) image factoring, which uses squared multiple correlations (SMCs) between each variable and all others as communalities to generate a mathematically determined solution with error variance and unique variance eliminated; (4) maximum likelihood factoring, which estimates the factor loadings for the population that maximizes the likelihood of sampling the observed correlation matrix (and has significance tests for factors); (5) alpha factoring, which maximizes the generalizibility of orthogonal factors and is somewhat likely to produce communalities greater than one; (6) unweighted least squares, which minimizes squared residual correlations; and (7) generalized least squares, which weights variables by shared variance before minimizing squared residual correlations (Tabachnick and Fidell 1989).

Other descriptions of these methods may be found in Johnson and Wichern (1988), and Mardia et al. (1979), or texts and manuals accompanying popular statistical packages.

4.8
Other Significance Tests – Comparisons of Solutions and Datasets

The comparison of both data sets and solutions can be done using a number of procedures described in Tabachnick and Fidell (1989). The procedures use Cattell's "salient similarity index" to compare patterns of factor loadings. The second method of comparison uses a correlation technique for looking at pairs of factor scores that are generated (see Tabachnick and Fidell 1989). It is also a good idea to plot the factor loadings of the factors against one another to ascertain if there is a need for oblique rotation, presence of another factor, unsuitable data, or any simple structure in the data.

4.9
Factor Analysis Procedure

The procedure in factor analysis is generalized in this section. Factor analysis is used to interpret the structure within a variance-covariance matrix of a multivariate data collection. The technique which it uses is the extraction of the eigenvalues and eigenvectors from the matrix of correlations, or covariances. The basic mathematics is given in most statistical texts (Davis 1973; Mardia et al. 1979; Johnson and Wichern 1988).

In general, suppose we compute a p-variate response vector [X] from a population which has mean, μ, and covariance matrix [σ]. The factor analysis model assumes there are m underlying factors (m < p variables) which we denote by f_1, f_2, ...f_m such that:

$$x_j = \mu_j + \lambda_{ji} f_1 + ... + \lambda_{jm} f_m + \sigma_j, \tag{4.2}$$

where $j = 1, 2, ..., p$;

and $\mathrm{Error}_{fk} = 0$, $\quad k = 1, 2, ..., m$, $\quad$ and $\mathrm{Error}\,\sigma_j = 0$, $j = 1, 2, ..., p$.

Thus, with appropriate assumptions, the factor model becomes a linear combination of underlying variables and specific error.

Factor analysis is used in many studies that require the analysis of processes that are related to a large number of mesasured variables in geology, geochemistry, and related subjects. Factor analysis will be applied in the next numerical examples to show some of its utility as a research tool. The important concepts to focus on in the analysis are factor loadings, communalities, and sum of explained variance.

4.10
Numerical Example on Water Chemistry

Factor analysis was used to study the chemistry of water in a coastal aquifer in Spain. Ruiz et al. (1990), completed an R-mode factor analysis that is applied to the hydrogeochemical study of a coastal aquifer located in Javea, Alicante

(Spain). A set of factors was found which explained the source of the ions, pH of the water in the aquifer, and certain chemical processes which accompany the intrusion of sea water such as strong sorption of potassium (K) by clay minerals. Table 4.2 shows the main factors and associated variance explained in the chemical data. Figure 4.2 is a depiction of the relations between the chemical constituents and the new factors defined in the factor analytic model. A factor loading, λ_{jk}, is defined as the weight or loading of the jth response variable on the

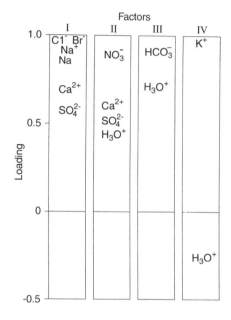

Figure 4.2. Diagrammatic representation of factor loadings from varimax rotation of chemical data for waters in Javea aquifer. (Modified from Ruiz et al. 1990) (Reprinted from Hydrology with kind permission of Elsevier Science-NL, Sara Burgerhart-straat 25, 1055 KV Amsterdam, The Netherlands. Copyright 1990).

Table 4.2. Varimax loading matrix and communalities for each analysed ion for waters in the Javea aquifer (Modified from Ruiz et al. 1990) (Reprinted from Hydrology with kind permission of Elsevier Science-NL, Sara Burgerhartstraat 25, 1055 KV Amsterdam, The Netherlands. Copyright 1990)

Variable	Factor 1	Factor 2	Factor 3	Factor 4	Comm.
Na	0.941	−0.007	−0.021	0.200	0.927
K	0.200	0.109	−0.008	0.937	0.931
Ca	0.676	0.598	0.136	−0.110	0.846
Mg	0.899	0.205	0.165	0.072	0.882
HCO_3	0.010	−0.240	0.903	0.120	0.887
Cl	0.969	0.159	0.016	0.078	0.971
SO_4	0.585	0.524	0.027	0.107	0.630
NO_3	0.090	0.918	−0.124	0.118	0.880
Br	0.965	0.148	0.054	0.045	0.960
H_3O	0.188	0.424	0.677	−0.353	0.800
Variance Explained %	51.1%	20.8%	15.4%	12.7%	
Cumulative %	51.1%	72.0%	87.4%	100.0%	

kth factor. The communality is defined as the portion or sum of variance explained by the common factors. The variance is equal to the square of λ_{jk}. The communality is then given as the sum of λ for each factor for that variable.

4.11
Numerical Example on Sediment Geochemistry

Factor analysis was used to investigate the geochemistry of sediments in Nigeria. Ekwere and Olade (1984) completed an R-mode factor analysis study to help locate factors accounting for element dispersion in granitic rocks of the Lituei Complex, younger granite province, Nigeria. They concluded from their study that five factors can account for 65% of the total data variance and that the granitic rocks are characterized by high values of Sn, Nb, Zn, Ta, Li, Rb, and F but ore-forming processes are related to postmagmatic changes. Table 4.3 shows the correlations among chemical constituents and Table 4.4 shows the interrelations among the factors and chemical constituents.

4.12
Numerical Example on Drainage Basin Properties

Factor analysis was used to investigate drainage basin properties in eastern Australia. The study (Abrahams 1972) looked at the morphometric data gathered from five fluvially eroded landscapes in Connors Range, Bobadah, Wyangala, Koetong, and Goulds Country in eastern Australia. Table 4.5 contains the final results. The five morphometric variables that were measured in all third order basins (except Wyangala) are: (1) basin area, A; (2) the cumulative channel length, L; (3) maximum basin relief, R; (4) basin perimeter, P; and (5) the aggregate stream number, N. The variables were made to be approximately normally distributed by taking logarithms. Principal component analysis was used to derive the initial factor matrix. Significant principal components were identified by an inspection of the matrix of the component loadings for each study area (Table 4.5). Those components were rejected that did not correlate significantly at the 0.05 significance level with any of the observed variables. In the Bobadah, Wyangala, and Goulds Country component matrices, three components were retained as significant. An examination of the corresponding eigenvalues indicates that together they account for 96–99% of the variance. Only two principal components were considered significant in the Connors Range and Koetong area component matrices. These two components account for 95% of the variation in the observed variables for each area. For the first factor, all variables are loaded heavily on this factor and are thought to be a result of geologic controls. The second factor is taken as a relief factor since only one variable is loaded on it. Factor three is believed to be an interaction of geologic and degradation or relief factors. The hypothesis is therefore generalized that factor 1 is a measure of the variance in the planimetric variables $A^{1/2}$, L, P, and N derived from the involvement of a stochastic link-generating process in the development of the drainage network in each of the landscapes under study (Abrahams 1972).

Table 4.3 Correlation coefficients, Ririwai biotite granite. (Modified from Ekwere and Olade 1984) (Reprinted from Hydrology with kind permission of Elsevier Science-NL, Sara Burgerhartstraat 25, 1055 KV Amsterdam, The Netherlands. Copyright 1984)

Variables	1	2	3	4	5	6	7	8	9	10	11	12	13	14
1 F	1.00													
2 Li	0.448	1.00												
3 Sn	0.035	0.354	1.00											
4 Rb	0.448	0.381	0.072	1.00										
5 Nb	0.018	0.458	0.355	0.255	1.00									
6 Zr	0.262	-0.121	0.187	0.041	0.206	1.00								
7 Pb	-0.059	0.230	0.035	0.110	0.179	0.131	1.00							
8 Cu	-0.115	-0.072	0.227	-0.141	0.163	0.087	0.174	1.00						
9 Zn	-0.015	0.483	-0.109	0.343	0.252	-0.237	0.173	0.102	1.00					
10 Ta	-0.020	0.504	0.220	0.168	0.560	0.186	0.134	0.222	0.382	1.00				
11 U	0.125	0.387	0.690	0.172	0.576	0.159	0.136	0.163	0.010	0.165	1.00			
12 La	0.112	0.066	0.106	0.102	-0.057	0.079	-0.146	0.019	-0.201	-0.108	0.073	1.00		
13 P	-0.208	0.184	0.576	0.047	0.290	0.067	-0.011	0.147	-0.134	0.003	-0.569	0.051	1.00	
14 Ti	0.115	-0.188	-0.556	0.071	-0.209	0.078	0.124	-0.203	0.183	-0.195	-0.503	0.174	-0.619	1.00

Table 4.4. R-mode varimax factor matrix, Ririwai biotite granite. (Modified from Ekwere and Olade 1984) (Reprinted from Hydrology with kind permission of Elsevier Science-NL, Sara Burgerhartstraat 25, 1055 KV Amsterdam, The Netherlands. Copyright 1984).

Factor 1: Sn, U, Sr, P, Cs, -Ti
Factor 2: F, Li, Zn
Factor 3: Sr, La
Factor 4: Nb, Zr, Ta
Factor 5: Pb, Cu

Factors	F1	F2	F3	F4	F5	Comm.
Variables						
F	−0.083	0.031	0.182	0.686	−0.394	0.667
Li	−0.272	0.806	0.118	−0.081	−0.084	0.751
Sn	0.755	0.122	0.261	0.137	0.127	0.688
Rb	−0.037	0.670	0.206	−0.058	−0.114	0.509
Sr	0.341	−0.061	0.779	0.132	−0.044	0.738
Nb	0.394	0.547	−0.013	0.350	0.302	0.668
Zr	0.062	−0.157	0.148	0.761	0.224	0.679
Pb	−0.110	0.253	0.049	−0.031	0.655	0.509
Cu	0.223	−0.209	−0.018	−0.064	0.722	0.619
Zn	−0.143	0.762	−0.317	−0.100	0.017	0.713
Ta	0.197	0.549	−0.328	0.383	0.226	0.707
U	0.715	0.279	0.327	0.173	0.110	0.738
La	−0.060	−0.006	0.759	0.015	−0.054	0.583
P	0.819	−0.025	0.191	−0.194	0.056	0.749
Ti	−0.902	0.085	0.209	0.083	0.057	0.876
Cs	0.857	0.076	−0.236	0.106	0.014	0.808
W	−0.059	0.093	0.829	0.328	0.136	0.836
Eigenvalues as %	25	16	14	8	6	
Cumulative %	25	41	55	63	69	

Table 4.5. Truncated principal components matrices, varimax rotated factor matrices, and communalities. (Modified from Abrahams 1972) (Reprinted by permission of Water Resources Research. Copyright 1972)

New variables	Truncated principal components initial factor matrix			Rotated factor matrix and communality (comm.)			
	C1	C2	C3	F1	F2	F3	Comm.
Goulds Country – three factors							
$\log A^{1/2}$	0.963	−0.064	−0.201	0.850	0.282	0.415	0.975
$\log L$	0.976	−0.170	0.058	0.719	0.207	0.651	0.984
$\log R$	0.616	0.786	0.051	0.229	0.959	0.165	0.999
$\log P$	0.952	−0.100	−0.254	0.884	0.240	0.378	0.981
$\log N$	0.903	−0.179	0.386	0.460	0.196	0.864	0.996
Eigenvalues	3.984	0.692	0.260				
Percent Variation	79.68	13.85	5.19	Total = 98.72			

Table 4.5 (continued)

New variables	Truncated principal components initial factor matrix			Rotated factor matrix and communality (comm.)			
	C1	C2	C3	F1	F2	F3	Comm.
Bobadah – three factors							
$\log A^{1/2}$	0.952	−0.221	−0.175	0.948	0.051	0.291	0.985
$\log L$	0.983	−0.063	0.029	0.834	0.158	0.500	0.971
$\log R$	0.359	0.922	−0.146	0.075	0.987	0.145	1.00
$\log P$	0.951	−0.198	−0.190	0.946	0.076	0.280	0.980
$\log N$	0.896	0.145	0.414	0.528	0.240	0.812	0.996
Eigenvalues	3.708	0.963	0.261				
Percent Variation	74.15	19.26	5.21	Total = 98.62			
Wyangala – three factors							
$\log A^{1/2}$	0.958	−0.040	−0.057	0.851	0.153	0.420	0.923
$\log L$	0.974	−0.064	−0.154	0.916	0.136	0.348	0.997
$\log R$	0.316	0.948	−0.027	0.100	0.992	0.070	0.999
$\log P$	0.805	−0.058	0.584	0.410	0.080	0.905	0.993
$\log N$	0.907	−0.167	−0.283	0.942	0.027	0.206	0.930
Eigenvalues	3.438	0.935	0.449				
Percent Variation	68.76	18.70	8.98	Total = 96.44			

Koetong – two factors				F1	F2		Comm.
$\log A^{1/2}$	0.970	0.045		0.748	0.620		0.943
$\log L$	0.971	−0.173		0.880	0.447		0.974
$\log R$	0.765	0.610		0.243	0.948		0.958
$\log P$	0.964	0.042		0.745	0.614		0.932
$\log N$	0.863	−0.443		0.996	0.166		0.941
Eigenvalues		4.144	0.060				
Percent Variation	82.88	12.05	Total = 94.93				

Connors Range – two factors				F1	F2		Comm.
$\log A^{1/2}$	0.976	−0.016		0.868	0.447		0.954
$\log L$	0.983	−0.101		0.914	0.375		0.976
$\log R$	0.623	0.766		0.187	0.969		0.974
$\log P$	0.967	0.043		0.832	0.495		0.938
$\log N$	0.844	−0.477		0.970	−0.022		0.941
Eigenvalues	3.955	0.827					
Percent Variation	79.10	16.53	Total = 95.63				

4.13
Numerical Example on Detrital Sediment Characteristics

Griffiths (1966) uniquely used factor analysis to look at geological process-variable relationships for detrital sediments. This early paper is one of several papers that showed the practical side of geostatistics and specifically that multivariate procedures are important geological tools. He concluded that variation in the properties of detrital sediments, as measured in thin section, contains much redundant information. The most obvious features which emerge are related to variation in size and size-sorting, indicating effects of the process of sorting, and variation in proportion of three cementation constituents which reflect diagenetic processes. This paper was very process-oriented and important to the development of multivariate procedures used today in the geosciences. Tables 4.6 and 4.7 contain the correlation matrix and components matrix for data from the example for the Cow Run sandstone. Table 4.8 contains the rotated matrix for the Cow Run sandstone data. It is readily apparent from the unrotated components matrix tables that porosity is related to processes controlling quartz, matrix, and grain length, and permeability is related to processes controlling rock fragments, matrix, and sorting.

Table 4.6. Correlation matrix for Cow Run sandstone. (Modified from Griffiths 1966) (Reprinted from Journal Geology, by kind permission of the University of Chicago Press, v. 14, Pages 653–671. Copyright 1966)

Variable	1	2	3	4	5	6	7	8	9	10
1 Quartz	1.00									
2 Rock frag.	−0.429	1.00								
3 Matrix	−0.245	−0.68	1.00							
4 Silica	0.225	−0.289	−0.15	1.00						
5 Size a	−0.30	−0.144	0.345	−0.075	1.00					
6 Size b	−0.29	−0.168	0.36	−0.072	0.994	1.00				
7 Sorting a	0.07	0.14	−0.41	0.29	−0.032	−0.064	1.00			
8 Sorting b	0.002	0.083	−0.250	0.291	−0.283	−0.312	0.829	1.00		
9 Porosity	0.399	−0.028	−0.200	0.265	−0.583	−0.572	0.281	0.088	1.00	
10 Perm.	0.544	−0.304	0.030	0.183	−0.528	−0.516	−0.424	−0.23	0.753	1.00

Values are significant if < or >.374, at the 5% significance level.

Explanation
1 Porportion of quartz   6 Grain size of b-axis (short)
2 Rock fragments   7 Sorting a (standard deviation)
3 Matrix   8 Sorting b (standard deviation)
4 Silica content   9 Porosity in %
5 Grain size of a-axis (long)   10 Permeability, in logarithm

Table 4.7. Components matrix for Cow Run sandstone. (Modified from Griffiths 1966) (Reprinted from Journal of Geology, v. 74, Pages 653–671, by kind permission of the University of Chicago Press. Copyright 1966)

Variables	Components					
	1	2	3	4	5	6
1 Quartz	−0.588	−0.190	−0.422	−0.342	−0.558	−0.025
2 Rock frags.	0.006	0.563	0.769	−0.228	0.134	0.066
3 Matrix	0.415	−0.628	−0.296	0.563	0.109	0.073
4 Silica cement	−0.303	0.138	−0.654	−0.310	0.566	−0.212
5 Mean length a	0.882	−0.149	−0.210	−0.340	0.015	0.183
6 Mean length b	0.879	−0.183	−0.211	−0.343	0.017	0.165
7 Sorting a	−0.003	0.860	−0.427	0.018	−0.140	0.112
8 Sorting b	−0.198	0.780	−0.393	0.347	0.026	0.206
9 Porosity	−0.809	−0.288	0.086	−0.146	0.261	0.348
10 Permeability	−0.757	−0.546	−0.045	−0.033	0.020	0.122
Eigenvalues accounted for	3.428	2.550	1.720	0.959	0.751	0.308
Cumulative %	34.27	59.77	76.97	85.56	94.07	97.15

Table 4.8. Rotated matrix of factor loadings for Cow Run sandstone. (Modified from Griffiths 1966) (Reprinted from Journal of Geology, v. 74, pages 653–671, with kind permission of the University of Chicago Press. Copyright 1966)

Variables	Components						
	1	2	3	4	5	6	Comm.
1 Quartz	−0.160	0.031	−0.029	−0.957	0.084	0.195	0.994
2 Rock frags.	−0.089	0.032	0.864	0.420	−0.224	−0.009	0.991
3 Matrix	0.183	−0.199	−0.917	−0.239	−0.121	−0.036	0.994
4 Silica	−0.011	0.201	−0.046	−0.109	0.961	0.144	0.999
5 Mean length a	0.952	−0.096	−0.127	0.106	−0.008	−0.225	0.997
6 Mean length b	0.948	−0.129	−0.143	0.097	0.000	−0.226	0.997
7 Sorting a	0.051	0.905	0.223	−0.111	0.145	−0.225	0.977
8 Sorting b	−0.218	0.947	0.020	0.076	0.113	−0.029	0.982
9 Porosity %	−0.383	−0.144	−0.194	−0.383	0.099	0.639	0.943

4.14
Summary

Factor analysis is used in the sciences to understand how different variables relate to one another and to the underlying controlling processes. Factors can often be defined as underlying theoretical variables.

4.15
Supplemental Reading

Johnson RA, Wichern DW (1988) Applied multivariate statistical analysis. Prentice-Hall, Englewood Cliffs

CHAPTER 5

Canonical Correlation

5.1
Concept

Canonical correlation is the analysis of the correlation between two sets of variables wherein the linear relationship between two sets of variables is maximized.

5.2
Definitions

Canonical Analysis. The most general form of multivariate techniques used in multiple regression, discriminant function analysis, and MANOVA.

Canonical Variates. These are linear combinations of variables; one set on the independent variable side and one set on the dependent variable side form a pair of canonical variates; hence the first reliable pair and second reliable pair form the first and second canonical variates.

Canonical Variate Scores. Are scores of subjects that were sampled.

Canonical Correlation Matrix (CCM). Is the product of four correlation matrices (submatrices), i.e., between dependent variables (inverted), between independent variables (inverted), and between dependent variables and independent variables. The correlation matrix is subdivided into four parts for analyses: Ryy, the dependent variable correlations; Rxx, the independent variable correlations; and correlations between Rxy and Ryx are correlations between independent and dependent variables.

5.3
Overview of Method

Canonical correlation is a generalization of the multiple correlation method. It is a method that solves the problem of simultaneously reducing two symmetric matrices to a diagonal form. The solution of the CCM for eigenvalues and eigenvectors is done in this procedure. Solving for the eigenvalues of a matrix is simply a redistribution process for the variance in the matrix, thus forming a few composite variates rather than many individual variables (Tabachnick and Fidell 1989).The relationship between a canonical correlation and an eigenvalue is: $\lambda_i = r_{ci}^2$, i.e., each eigenvalue is equal to the squared canonical correlation for the pair of canonical variates. Hence the canonical correlation equals the square root of the eigenvalue, and is interpreted as the Pearson product-moment correlation. When squared, the r_{ci} (canonical correlation) is the common overlapping variance between two variables.

Variables are often found to belong to different groupings that are related to different processes or factors. Canonical correlation analyses are used to identify and quantify the associations between two sets of variables in a data set. Canonical correlation analyses have the objective of determination of the correlation between a linear combination of the variables in one set and a linear combination of the variables in another set. The first pair of linear combinations have the largest (maximum) correlation. The second pair of linear combinations are determined and have the second largest correlation of the remaining variable sets. This process continues until all pairs of remaining variables are analyzed. The pairs of linear combinations are called canonical variables, and their correlations are called canonical correlations. These parameters are evaluated to ascertain differences or similiarities between variable groups.

A canonical correlation analysis between two sets of variables yields one or several canonical correlation coefficients that index the degree of correlation between the two sets rather than the individual variables. The canonical correlation analysis seeks to define the maximum linear combination between linear combinations for each set of variables. Each linear combination is a canonical variate. The relationship between a pair of variates yields a canonical correlation coefficient. Each variate has a set of weights indicating the relative relation of each variable in the set to the variate. It is possible to test the statistical significance of canonical variates relative to the degree that linear relations between the two variable sets have been extracted. This involves the use of Wilks' λ and its test of significance by a chi-square method. One may loosely think of canonical variates as factors, because a squared canonical correlation coefficient, like the sum of squared loadings on a factor, is called an eigenvalue. Finally, the variable weightings on a canonical variate are interpreted loosely akin to factor loadings.

Various similarities exist between canonical correlation and other multivariate methods. In multiple discriminant analysis, the degree of correlation between the discriminating variables and the groups is called the canonical correlation for that discriminating function. Canonical correlation analysis may also be used on qualitative data such as contingency tables (see Johnson and Wichern 1988). Correlation between a set of X variables and a set of Y variables is called canonical correlation and the technique is related in principle to multiple regression, but can take into account multiple dependent variables.

5.4
Significance Tests – Importance of Wilks' Lambda (λ)

When variables are considered individually, λ is the ratio of the within-groups sum of squares to the total sum of squares. A λ of one occurs when all observed group means are equal. Values close to zero occur when within-groups variabi-lity is small compared to the total variability, i.e., when most of the total variabi-lity is attributable to differences between the means of groups. Thus, large values of λ indicate that group means do not appear to be different, while small values indicate that group means do appear to be different (Norusis 1985). In the case of multiple samples where there are no differences between groups, a test that the means of all discriminant functions in all groups are equal to zero is done with a Wilks' λ. Wilks' λ is the product of the univariate Wilks' λ for each func-tion and is not just the ratio of the between-groups to within-groups sums of squares. The significance of observed Wilks' λ is then based on a chi-square transformation to the statistic (see Norusis 1985).

5.5
Significance Tests – Bartlett's Test Using Chi-Square Variable

The significance test for determining whether one or a set of r_{ci}'s (canonical correlations) differ from zero is done through a chi-square test, where N = no. of cases, kx = No. of variables in set one of independent variables, and ky is the number of variables in the dependent variables set, with ($\lambda(m) = Lm$) natural log Lm = the product of differences between eigenvalues and unity, generated across m canonical correlations (see Tabachnick and Fidell 1989). The degrees of freedom = kx times ky for test. Large chi-square (small probability of < 0.05) signifies that a significant overlap in variability exists between two sets and a reliable relationship does exist. Many other questions can be answered using the loading matrix for the data set, i.e., how much variance is partitioned within canonical variates set and between sets. Each set of canonical variates is inter-preted separately.

5.6
Canonical Correlation Procedure

Canonical correlation is a generalization of multiple correlation when groups of similiar variables exist, i.e., q variables in vector X_1 and p – q variables in vector X_2. If an investigator is interested in the relationship of these two sets when they are not independent, the interrelation may be almost completely described by the correlation between a few linear combinations of the responses in each vector. If q = 1, the multiple correlation coefficient is the largest correlation attainable between X_1 and a linear combination of the components in vector X_2. When q > 1, a natural generalization is to find the largest correlation attainable between a linear combination of vector X_1 and linear combination of vector X_2. Thus we want to find the maximum correlation between two sets, and the maxi-mum correlation is ϱ_1, which is called the "first canonical correlation" between vector X_1 and vector X_2, and the determined U_1 and V_1 variables are called the "first canonical variates". The square of ϱ_1 is taken as the first eigenvalue. The next maximum correlation ϱ_2 is determined for the next two sets of variables, as is U_2 and V_2, the "second canonical variates", until all sets of variables are exhausted.

Fig. 5.1. Plot of transformed scores for first canonical correlation for soil data. (Ratha and Sahu 1993) (Reprinted from Environmental Geology with permission of Springer-Verlag. Copyright 1993)

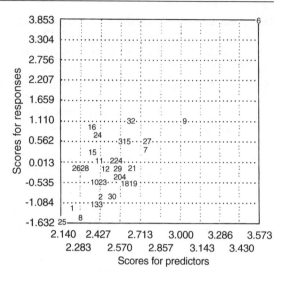

Fig. 5.2. Plot of canonical transformed scores for second canonical correlation for soil data. (Ratha and Sahu 1993) (Reprinted from Environmental Geology with permission of Springer-Verlag. Copyright 1993)

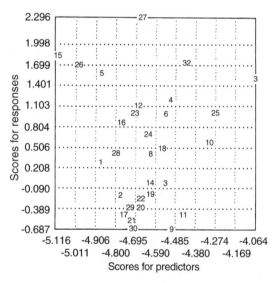

5.7
Numerical Example on Sediment and Soil Pollution

Canonical correlation was used to investigate geologic variables related to soil and sediment pollution in Bombay, India.

Ratha and Sahu (1993) uniquely applied canonical correlation to the analysis of pollutants in soils and sediments of the Bombay region, India. Few applications of this procedure are found in the geoscience literature and for this reason, this study is important. They concluded from the study of soil samples, that the first canonical correlation explains the role of mean particle size

Fig. 5.3. Plot of canonical transformed scores for the first canonical correlation for sediment. (Ratha and Sahu 1993) (Reprinted from Environmental Geology with permission of Springer-Verlag. Copyright 1993)

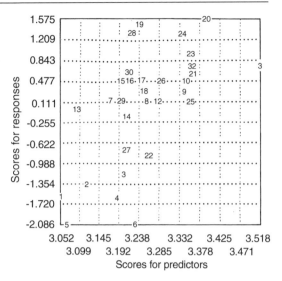

Fig 5.4. Plot of canonical transformed scores for the second canonical correlation. (Ratha and Sahu 1993) (Reprinted from Environmental Geology with permission of Springer-Verlag. Copyright 1993)

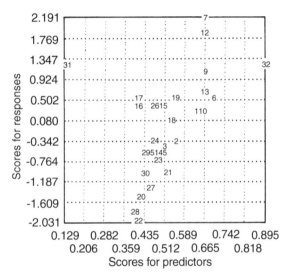

in controlling the concentration of *Pb* and *Ni*. The second correlation shows the role of clay minerals in controlling the concentration of trace elements such as *Fe* and *Zn*. Plots of the transformed scores of the first canonical correlation illustrate that there is a clear distinction between the type of sediments collected from Thane Creek and the Ulnas River Region (Figs. 5.1, 5.2, 5.3, and 5.4).

Tables 5.1 and 5.2 show the correlation coefficients of normalized variables, and Tables 5.3 and 5.4 show the correlation between original variables and canonical variates.

Tables 5.1. Correlation coefficient of normalized physical and chemical parameters of sediments. (Modified from Ratha and Sahu 1993) (Reprinted from Environmental Geology with permission of Springer-Verlag. Copyright 1993)

Parameters	1	2	3	4	5	6	7	8	9	10	11	12	13
1 Cd	1.00												
2 Co	-0.349	1.00											
3 Cr	0.004	-0.045	1.00										
4 Cu	0.422	-0.328	-0.092	1.00									
5 Fe	0.659	-0.596	0.302	0.581	1.00								
6 Mn	0.568	-0.699	0.217	0.651	0.885	1.00							
7 Ni	0.272	-0.327	0.075	0.704	0.531	0.606	1.00						
8 Pb	0.117	-0.052	-0.534	0.671	0.007	0.119	0.424	1.00					
9 Zn	0.437	-0.411	0.080	0.612	0.541	0.533	0.274	0.275	1.00				
10 Mean	0.04	-0.061	-0.476	0.461	-0.089	-0.009	0.199	0.820	0.088	1.00			
11 Sd	-0.193	0.026	-0.403	0.229	-0.229	-0.102	0.151	0.478	-0.129	0.505	1.00		
12 Sk	-0.180	0.167	0.363	-0.445	0.017	-0.119	-0.239	-0.532	-0.383	-0.63	-0.349	1.00	
13 Ku	0.041	-0.065	-0.021	0.014	0.122	0.115	0.084	0.088	-0.075	0.28	0.041	-0.129	1.00

Note: Sd, Standard deviation; Sk, skewness; Ku, kurtosis

Tables 5.2. Correlation coefficient of normalized physical and chemical parameters of soil (After Ratha and Sahu, 1993) (Reprinted from Environmental Geology with permission of Springer-Verlag. Copyright 1993)

Parameters	1	2	3	4	5	6	7	8	9	10	11	12	13
1 Cd	1.00												
2 Co	0.628	1.00											
3 Cr	0.462	0.452	1.00										
4 Cu	0.016	0.132	0.018	1.00									
5 Fe	0.236	0.231	0.307	0.444	1.00								
6 Mn	0.495	0.823	0.539	0.249	0.244	1.00							
7 Ni	0.669	0.755	0.611	0.121	0.284	0.681	1.00						
8 Pb	0.382	0.311	0.299	0.098	0.245	0.223	0.244	1.00					
9 Zn	0.232	0.460	-0.210	-0.154	-0.074	0.165	0.076	0.064	1.00				
10 Mean	0.045	-0.091	-0.366	0.026	-0.218	-0.291	0.006	0.015	0.214	1.00			
11 Sd	0.279	0.245	0.282	0.117	0.464	0.225	0.042	0.266	0.164	-0.530	1.00		
12 SK	-0.025	-0.022	0.250	-0.037	-0.019	0.166	0.060	-0.315	-0.245	-0.65	0.308	1.00	
13 Ku	0.044	0.008	-0.191	-0.002	-0.343	-0.001	0.218	-0.036	-0.031	0.367	-0.365	-0.299	1.00

Sd, Standard deviation; Sk, skewness; Ku, kurtosis

Table 5.3. Correlation between original variables and derived canonical variates for soils. (Modified from Ratha and Sahu 1993) (Reprinted from Enviromental Geology with permission of Springer-Verlag. Copyright 1993)

	Responses	Rdx	Predictors	Rdy	C.S.	DF
Rc1 = 0.84	−0.575 Co	0.069	−0.616 SD	0.203	49.77	36
	−0.700 Fe		0.649 Sk			
	1.122 Ni		0.555 Ku			
Rc2 = 0.65	−0.455 Cr	0.038	0.998 Mean	0.147	19.65	24
	−0.713 Mn					
	0.603 Ni					
	0.523 Zn					

Df, degrees of freedom; Sd, standard deviation; Sk, skewness; Ku, kurtosis; C.S., chi-square; Rc1, first canonical variate correlation; Rc2, second canonical variate correlation

Table 5.4. Correlation between original variables and derived variates for sediment. (Modified from Ratha and Sahu 1993) (Reprinted from Enviromental Geology with permission of Springer-Verlag. Copyright 1993)

	Responses	Rdx	Predictors	Rdy	C.S.	DF
Rc1 = 0.92	1.008 Pb	0.084	1.163 Mean	0.183	75.01	36
	−0.439 Ni					
Rc2 = 0.77	1.089 Fe	0.071	1.182 Sk	0.147	19.65	24
	−0.644 Mn					

Df, degrees of freedom; Sk, skewness; C.S., chi-square; Rc1, first canonical variate correlation; Rc2, second canonical variable correlation

5.8
Summary

Canonical correlation analysis is very important in analyzing groups of contrasting variables. Few geohydrologic studies have been completed using the canonical correlation method.

5.9
Supplemental Reading

Johnson RA, Wichern DW, (1988) Applied multivariate statistical analysis. Prentice Hall, Englewood Cliffs

CHAPTER 6

Multiple Regression

6.1
Concept

The objective of multiple regression is to derive a linear relationship between a dependent variable and a group of independent variables.

6.2
Definitions

General Linear Model. Mathematical form of the equation used to predict a dependent variable from a group of independent variables.

Residuals. Identifies outliers in the solution, and can be used to detect failure of normality, nonlinearity, and heteroscedasticity when plotted against predicted y'.

Leverage. A distance measure that identifies outliers in the independent variables and may be denoted as a variation of the diagonal elements of a "hat" matrix.

Influence. A distance measure that identifies cases that are too influential and is a measure of the change in regression coefficients produced by leaving out a case.

Ridge Regression. A method of regression used when all the independent variables are wanted in the equation, even though multicollinearity has been detected.

6.3
Overview of Methods

Quite often, the methods for estimating one variable by means of a related variable yield poor results, not because the relationship is far removed from the assumed linear one but because there is no single variable sufficiently closely related to the variable being estimated to yield good results. However, in this instance it may be that there are several variables which, when taken jointly, will serve as a satisfactory basis for estimating the desired variable. Since linear functions are simple to manipulate and experience shows that many variables are approximately linearly related, it is reasonable to attempt to estimate the desired variable by means of a linear function of the remaining variables. The next section describes these models.

This section deals with the general linear model, which is in contrast to the simple linear model. The simple linear model is the equation of a straight line, $y = a + bx$, whereas the general linear model is of the mathematical form:

$$y = \beta_0 + \beta_1 x_1 + \beta_2 x_2 + \ldots + \beta_k x_k \tag{6.1}$$

where $x_1, x_2, \ldots x_k$ may be different variables or functions of other variables. It is called the general linear model because the individual terms are added together. The individual terms may represent variables raised to powers of x, such as x^2 or x^3, or products of x, such as $x_1 x_2$. The βs are the regression coefficients that are estimated during the fitting of the general linear model.

Multiple-regression analysis is used to derive an equation that can be used to predict values of the dependent variable, y, from several independent variables. It thus provides the framework for determining if a variable x_i makes a significant contribution to the prediction of y, i.e., is the coefficient = zero. If so found, the variable is deleted from the prediction model, thus simplifying the model. The problem narrows down to finding the best function of the form represented by the equation given below to predict the mean value of y from the x's, and is often done by least squares estimation which minimizes the residual sum of squares that is done through the application of the normal equations. This procedure leads to an analysis of variance test of significance for the relationship between y and the x's. Any observed variable can be considered to be a function of any other variables measured on the same samples (Davis 1973). The steps involved in model building can include some or all of the following: (1) model equation setup, (2) testing of hypotheses, (3) overall test of the model, (4) individual tests of coefficients, (5) a partial test (F-tests of groups of coefficients simultaneously) of the model, and (6) determination of confidence intervals for coefficients.

6.4
Multiple Regression Procedure

In multiple regression, the equation is of the general form:

$$y = b_0 + b_1 X_1 + b_2 X_2 + \ldots + b_n X_n + E, \tag{6.2}$$

where b_0 to bn are partial regression coefficients, and X_1 to X_n are measured variables and E is the error term. The B's in a multiple regression model are estimated by b's, the sample partial regression coefficients, as in Eq. 6.2. They are called partial regression coefficients because each gives the rate of change (or slope) in the dependent variable for a unit change in that particular variable, provided all other independent variables are held constant (Davis 1973). For example, this could lead to notation of the form:

$$Y = b_0 + b_{1.23} X_1 + b_{2.13} X_2 + b_{3.12} X_3 + E, \tag{6.3}$$

where the coefficient $b_{1.23}$ is defined as the regression coefficient of variable 1 on Y, as variables, 2 and 3 remain constant.

The null hypothesis and alternative hypothesis for the regression take the following form:

H_0: There is no linear relationship between the dependent variable Y and the set of independent variables X_1 to X_n.

H_1: There is a relationship between Y and the set of independent variables X_1 to X_n.

The F-test and the analysis of variance give the overall significance of the regression test (Minitab Inc., 1986). The coefficient of multiple determination, R^2, provides an overall measure of the adequacy of the equation to predict. The coefficient of multiple determination (coefficient of determination for simple regression) is a measure of how well the independent variables can explain changes in the dependent variable; its value is between zero (meaning poor fit) and one (meaning perfect fit). We can evaluate the importance of the individual explanatory variables by examining the sample partial regression coefficients associated with each (Daniel and Terrell 1983). The setting up of the general form of the equation used to predict is the result of a regression analysis. The F-statistic is used to test the hypothesis that the true value of each coefficient in a regression equation is zero. The t-statistic is used to test the hypothesis that the true value of one specific coefficient is zero.

6.5
Stepwise Multiple Regression

In many studies of multivariate data, a different approach to the analysis of the data may give a different perspective and result. To further reduce the required number of measured variables and to analyze and remove collinearity that might exist between variables, another analytical technique, such as stepwise regression, is often applied to data. Stepwise regression uses the same analytical optimization procedure as multiple regression but differs from the multiple-regression method in that a subset of predictor variables is selected sequentially from a group of predictors by means of statistical testing of hypotheses.

The parameters of concern in this analysis are the T-test and its corresponding probability. The stepwise multiple regression technique is one way to deal with the multicollinearity problem in regression analyses.

6.6
Significance Tests – Addressing Multicollinearity

Multicollinearity in multiple regression analysis can be a serious problem. Several ways of addressing this problem are available. Among the test statistics are:

1. The variance inflation factor (VIF) for variable, X_j, where:

$$VIF = 1/(1 - R_j^2) \tag{6.4}$$

for X_j regressed on remaining $k-1$ predictors; and $R_j^2 =$ coefficient of determination for this regression:

2. the Durbin-Watson statistic for serial correlation;
3. the pure error lack of fit test; and
4. the experimental lack of fit test.

These tests can routinely be done when using software packages such as Minitab (1986), SAS, SPSS, or others noted in earlier discussions.

6.7
Multiple Regression Applications

Many unique applications of multiple regression exist in the literature, but in keeping with the theme and continuity of this book, some insight and unique applications of these procedures are provided.

A comprehensive summary of multiple regression is found in Draper and Smith (1981), but an abbreviated discussion is helpful here. As stated earlier, a simple linear model is based on the equation of a straght line, but in multiple regression, the mathematical equation is of the general linear model form:

$$y = b_0 + b_1X_1 + b_2X_2 + \ldots + b_nX_n + E, \tag{6.5}$$

where b_0 to b_n are partial regression coefficients, and X_1 to X_n are different variables. As also previously stated, the model is called a general linear model because the individual terms are additive. The goal of the analysis is to fit the general model to the data. In regression analyses, variables may act jointly giving what is called an interaction and is handled using cross-products. Their usefulness is usually detected through a graphically-oriented process.

The null hypothesis and alternative hypothesis for the regression are:

H_0: There is no linear relationship between Y and the set of independent variables X_1 to X_n.
H_1: There is a relationship between Y and the set of independent variables X_1 to X_n.

The F-test and the analysis of variance give the overall significance of the regression test. The coefficient of multiple determination, R^2, provides an overall measure of the adequacy of the equation to predict the dependent variable from other independent variables. The coefficient of multiple determination (coefficient of determination for simple regression) is a measure of how well the independent variables can explain changes in the dependent variable with its value lying between zero (meaning poor fit) and one (meaning perfect fit). The lack of fit of a regression model or adequacy can be easily examined through the analysis of residuals which can be plotted against the predicted variables, or against time if the data is chronological (the Durbin-Watson statistic can also be applied in this instance). Other plots such as histograms or normal probability plots can be used to detect outliers or severe departures from normality. It is possible to further evaluate the importance of the individual explanatory variables by examining the sample partial regression coefficients associated with each (Daniel and Terrell 1983). The formulation of the general form of the equation to predict is the result of the regression analysis and an F-statistic is used to test the hypothesis that the true value of each coefficient in a regression equation is zero. Likewise, the T-statistic is used to test the hypothesis that the true value of one specific coefficient is zero. The Press selection method as explained in Draper and Smith (1981) is another method for looking at the proper selection of variables in the predictive equation, but the computational requirements are sometimes enormous and may be a drawback. Draper and Smith (1981) explained in detail methods such as the R-squared statistic, least-squares versus ridge regression, principal components regression, and other methods for determining the best regression equation. This will not be done here.

6.8
Numerical Example on Water Yield

Multiple regression was applied in a hydrologic analysis of water yield to compare multivariate methods. McCuen et al. (1979) did a comparative study of statistical methods for water supply forecasting. They concluded that while the correlation coefficient and standard error of estimate are frequently used when comparing models of seasonal water yield, the criteria of most importance are rationality of regression coefficients, the distribution of residual errors, and the correctness of indicators of the relative importance of the predictor variable. This paper is an important systematic evaluation of variables using pertinent multivariate statistical procedures and their application to investigate aspects of water-supply forecasting. Tables 6.1 and 6.2 give the comparative results from the study.

Table 6.1. (A) Model parameters and (B) model equations for predicting April–July streamflow. (Modified from McCuen et al. 1979) (Reprinted from Water Resources Bulletin, vol. 15, No. 4, with permission of American Water Recources Association. Copyright 1979)

(A) Model parameters

Model	Method	R	R^2	S.E.E	Model
1	MR	0.887	0.787	12.22	$Q = -16.828 + 2.518\,x_5$
2	MR	0.906	0.820	12.71	See below (B)
3	SR	0.924	0.853	10.62	
4	PCRA $_2$	0.889	0.790	13.74	
5	PPC $_1$	0.882	0.778	13.07	
6	PS	0.897	0.804	13.35	
7	PS	0.895	0.801	13.01	
8	MR	0.960	0.921	7.62	

MR multiple regression; Q April–July streamflow ($\times 10^3$ acre-feet); R correlation coefficient; SR stepwise regression; R^2 coefficient of determination; PCRA principal components regression analysis with two principal components; PPC polynomial fitting using PCRA; PS pattern search numerical optimization; S.E.E. standard error of estimate.

Table 6.1. (continued)

(B) Model equations

Model	equation
2	$Q = -9.308 + 2.321\,x_1 - 1.283\,x_2 + 1.226\,x_3 + 1.011\,x_4 + 1.449\,x_5 - 0.094\,x_6$
3	$Q = -22.14 + 6.902\,x_8 - 1.400\,x_7 + 4.013\,x_3$
4	$Q = -12.97 + 2.946\,x_1 + 0.942\,x_2 + 0.569\,x_3 + 0.713\,x_4 + 0.535\,x_5 + 1.366\,x_6$
5	$Q = 6.265 + 0.838\,x_5 + 0.0153\,x_5^2 + 0.0003213\,x_5^3$
6	$Q = 0.965\,x_1 + 1.618\,x_3 + 0.499\,x_4 + 0.424\,x_5 + 0.449\,x_6$
7	$Q = -4.067 + 1.324\,x3 + 0.290\,x_4 + 0.998\,x_5 + 0.381\,x_6$
8	$Q = -27.834 + 4.905\,x_8 + 3.493\,x_9$

x_1 = October baseflow (x10^3 acre-feet)
x_2 = April 1 snow water equivalent: Castle Valley (inches)
x_3 = April 1 snow water equivalent: Duck Creek (inches)
x_4 = April 1 snow water equivalent: Harris Flat (inches)
x_5 = April 1 snow water equivalent: Midway Valley (inches)
x_6 = April 1 snow water equivalent: Panguitch Lake (inches)
x_7 = March 1 snow water equivalent: Duck Creek (inches)
x_8 = December baseflow (x10^3 acre-feet)
$x_9 = (x_3 + x_4 + x_5)/3$ (inches)

Table 6.2. Residual errors for models. (Modified from McCuen et al. 1979) (Reprinted from Water Resouces Bulletin, vol. 15, No. 4, with permission of American Water Resources Association. Copyright 1979)

Year	Observed S.F. (April to July) (1 000 acre-feet)	Residual error (percent of observed flow) Model							
		1	2	3	4	5	6	7	8
1959	13.8	24.0	97.8	−90.6	78.9	53.3	83.7	−49.9	−44.0
1974	16.6	26.1	48.9	−48.8	35.8	40.7	33.7	−16.4	−27.3
1953	16.7	29.9	41.1	−52.2	66.0	42.6	49.0	−31.5	0.7
1960	17.4	49.3	24.4	5.7	38.5	52.3	52.1	−51.1	25.3
1955	17.5	77.3	95.0	−63.4	105.5	70.9	106.6	−95.4	70.9
1961	18.2	58.0	24.2	−48.3	57.0	55.7	56.6	−56.8	−4.1
1963	18.9	−71.8	−78.5	58.2	−47.6	−20.4	−47.4	63.3	44.4
1972	20.9	15.9	−18.5	50.0	−35.7	21.4	−34.8	29.5	−15.3
1956	21.0	52.5	16.9	20.5	−7.7	45.8	25.0	−32.8	13.5
1970	22.9	−10.7	−3.9	−42.4	15.7	0.7	−11.5	26.1	−13.8
1971	24.3	10.0	5.8	21.8	18.1	11.1	17.1	−12.7	8.5
1976	25.2	44.1	46.6	−48.8	69.0	34.0	67.9	−61.9	−53.3
1964	29.1	−44.5	−37.7	13.1	−24.6	−29.2	−14.7	23.7	23.6
1954	35.4	54.5	48.8	−11.9	38.2	40.5	40.1	−44.6	−23.4
1975	37.4	11.2	−0.4	−5.3	4.9	1.4	9.1	−10.7	−18.8
1966	39.4	5.6	4.6	−12.2	−15.8	−3.7	−5.6	4.6	9.5
1957	40.2	−4.7	0.3	4.5	−21.9	−12.1	4.0	−3.6	−4.4

Table 6.2. (continued)

Year	Observed S.F. (April to July) (1 000 acre-feet)	Residual error (percent of observed flow) Model							
		1	2	3	4	5	6	7	8
1965	51.9	−35.9	−33.2	−32.8	−36.6	−39.3	−27.1	28.7	14.9
1962	53.8	23.2	21.3	−27.3	19.9	15.4	24.0	−26.0	−0.8
1967	55.7	−41.1	−43.5	29.4	−54.3	−44.1	−47.7	47.9	26.1
1968	56.4	−17.8	−3.6	8.0	−4.7	−25.5	−6.4	11.3	7.4
1958	62.8	−5.3	−13.1	8.3	−0.4	−13.0	−10.9	9.2	−3.1
1952	88.0	13.7	8.4	−10.8	4.5	25.2	3.6	−9.1	−9.1
1973	91.3	−21.3	−18.5	18.7	−19.0	−24.7	−22.9	22.1	16.4
1969	107.1	−16.9	−8.8	8.4	−8.5	−13.9	−11.2	11.9	1.0

To convert acre-feet to ha-m, multiply by 0.123
Observed S.F. = Observed streamflow

6.9
Numerical Example on Streamflows

Multiple Regression was applied in a analyzing stream low flows. Tasker (1972) used regression analysis to assess the geohydrologic influence of geologic variables on estimates of low flow. He concluded that a ground-water availability factor in addition to drainage area extends a pronounced influence upon the areal variation in low flow characteristics of streams in southeastern Massachusetts. Tasker (1972) found that the annual minimum 7-day mean flow at the 2- and 10-year recurrence intervals (Q_2 and Q_{10}) are significantly related, with a standard error of 50 and 70%, respectively, to the drainage area and the average ground water available from wells in the basin. The regression equations derived are of the form:

$$\log(Q + 0.1) = a + b_1 \log A + b_2 G,\qquad\qquad(6.5)$$

where:

Q is Q_2 or Q_{10}, a is the regression constant, b_1 is the regression coefficient for A, b_2 is the regression coefficient of G, A is the drainage, and G is the ground-water factor and is a generalization of the geologic influence on low flows as defined in terms of well yield in specific parts of a ground-water basin (see Tasker 1972). Tables 6.3 and 6.4 summarize the findings from the analyses of low flow estimates. Figures 6.1 and 6.2 show the relations between drainage area and discharge.

Table 6.3. Drainage area, annual minimum 7-day mean flow at 2- and 10-year recurrence intervals, and ground-water factor of SE Massachusetts – partial listing. (Modified from Tasker 1972)

Station	Annual minimum 7-day mean flow (cfs)							
	2-year low-flow recurrence				10-year low flow recurrence			
	Variables[a]							
	1	2	3	4	5	6	7	8
Crooked Meadow	5.00	0.60	0.53	0.07	0.20	0.22	−0.02	92
Bound Brook	4.88	0.10	0.46	−0.36	0	0.17	−0.17	73
Indian Head Brook	4.39	0.50	0.72	−0.22	0.20	0.37	−0.17	170
Pudding Brook	4.53	0.50	0.63	−0.13	0.30	0.30	0	140
Third Herring Brook	9.85	0.40	1.31	−0.91	0	0.55	−0.55	117
Second Herring Brook	1.72	0.30	0.20	0.10	0	0.05	−0.05	29
First Herring Brook	1.72	0.02	0.06	−0.04	0	0	0	33
South River	7.48	2.00	0.86	1.14	1.50	0.35	1.15	95
Jones River Brook	3.57	0.90	0.84	0.06	0.60	0.50	0.10	236
Halls Brook	4.12	1.60	0.94	0.66	1.20	0.54	0.66	224

[a] cfs, cubic feet per second, except variables 1 and 8

Variables : 1 Drainage Area A (in sq. miler); 2. Estimated from baseflow; 3 Estimated from Eq. 6.5; 4. Residuals(variables estimated from computed values); 5. Estimated from baseflow; 6. Computed from Eq. 6.5; 7. Residuals: 8. Ground-water factor, G.

Table 6.4 Summary of regression analysis. [Modified from Tasker 1972); (see Eq. (6.5)]

| Dependent variable | Constants | | | Standard error | |
| | Regression coefficients | | Constant | Percent | Change | |
	for A | (b₁)	for G | (b₂)	(a)		
Q_2	1.39	0.0022	−1.030	105	...
	1.00	0.0036	−1.101	55	50
Q_{10}	...	0.0036	−0.806	105	...
	0.83	0.0028	−1.340	70	35

6.10
Numerical Example on Geomorphic Variables

Multiple regression was applied in the analysis of water yield in watersheds in Kentucky. Haan and Allen (1972) compared multiple regression and regression on principal components of data matrices containing 21 geomorphic variables. They analyzed water yield data from 13 small agricultural watersheds in Kentucky by standard multiple regression techniques. They used measured data

Fig. 6.1. Relation between $Q_2 + 0.1$, drainage area, and ground-water factor, where Q_2 is the annual minimum 7-day mean flow at the 2-year recurrence interval. (Tasker 1972)

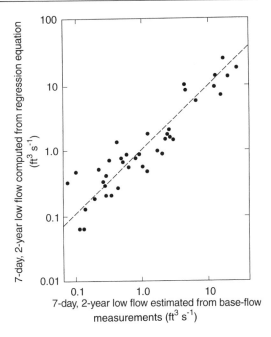

Fig. 6.2. Relation between Q_2 computed from regression equation and Q_2 estimated from baseflow measurements. (Tasker 1972)

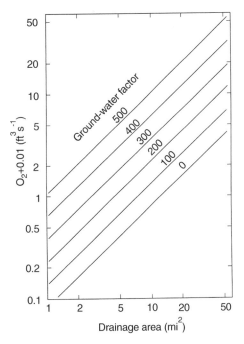

on 21 geomorphic variables, which they used to ascertain that multiple regression was a labor-saving method when evaluated against regression on principal components. Tables 6.5–6.8 show the correlation matrix and factor loadings and apparently support the conclusion that a maximum of three of the principal components are significant for this data.

Table 6.5. Correlation matrix. (Modified from Haan and Allen 1972) (Reprinted with permission from Water Resources Research, vol. 8, No. 9. Copyright 1972)

Variables	Variables[a]							
	A	S	L	P	di	Rs	F	Rr
A	1.00							
S	−0.17	1.00						
L	0.90	−0.21	1.00					
P	0.96	−0.10	0.92					
di	0.91	−0.16	0.67	0.81	1.00			
Rs	−0.25	−0.05	−0.58	−0.41	0.15	1.00		
F	−0.48	−0.30	−0.53	−0.48	−0.32	0.29	1.00	
Rr	−0.52	0.80	−0.54	−0.51	0.50	0.18	−0.08	1.00

[a] Geomorphic variables: A = area in sq. miles; S = average land slope, in percent; L = axial length, miles; P = perimeter, miles; di = diamater of largest circle that can be drawn in watershed,miles; Rs = dimensionless shape factor; F = stream frequency, streams/sq.miles; Rr = relief ratio, ft/miles

Table 6.6. Regression equations. (Modified from Haan and Allen 1972) (Reprinted with permission from Water Resources Research, vol. 8, No. 9. Copyright 1972)

Equation no.	Expression	R^2	S.E.	Overall F
(2)	Ro = −14.77* + 0.45* Pr + 0.17 A − 0.02 S + 0.29 L + 0.99*P −3.02 di + 5.64 Rs + 0.37 F + 0.013*Rr	0.97	0.69	10.0
(3)	Ro = −9.65* + 0.43* Pr + 0.62* P + 0.010* Rr	0.91	0.69	28.7
(4)	Ro = −4.24 + 0.51* Z_1 + 0.77* Z_2 −0.037 Z_3 + 1.11* Z_4 + 0.97 Z_5 −0.67 Z_6 + 3.41* Z_7 + 0.66 Z_8 + 0.45 Pr	0.97	0.69	10.0
(5)	Ro = −7.60 + 0.54* Z_1 + 0.76* Z_2 + 1.17* Z_4 + 3.44* Z_7 + 0.52* Pr	0.94	0.63	20.9
(6)	Ro = 4.24 + 2.11* Z_1 + 1.18* Z_2 + 0.69 Z_3 + 1.17* Z_4 + 1.69 Z_5 + 0.36 Z_6 + 3.61* Z_7 + 0.76 Z_8 + 0 .45* Pr	0.97	0.69	10.0
(7)	Ro = −1.19* + 1.48* Z_1 + 1.44* Z_2 + 1.06* Z_4 + 2.65* Z_7 + 0.62* Pr	0.91	0.75	14.7

Explanation: Ro, mean annual runoff; Pr, mean annual precipitation; Zi, principal component i.

S.E., standard error

* significant at 5% level

Table 6.7. Factor loadings of the principal components. (Modified from Haan and Allen 1972) (Reprinted with permission from Water Resources Research, vol. 8, No. 9. Copyright 1972)

Variables	Principal Components							
	1	2	3	4	5	6	7	8
A	0.967	0.077	0.191	0.087	0.101	−0.036	−0.043	0.043
S	−0.288	0.879	0.193	0.260	−0.192	0.041	−0.024	0.007
L	0.953	0.119	−0.200	0.040	0.073	0.175	0.018	−0.006
P	0.962	0.137	0.035	0.181	−0.050	−0.076	0.117	−0.006
di	0.823	−0.053	0.554	0.071	0.014	−0.029	−0.071	−0.034
Rs	−0.398	−0.227	0.872	−0.150	−0.003	0.062	0.057	0.011
F	−0.501	−0.668	0.027	0.547	0.035	0.024	−0.001	0.001
Rr	−0.648	0.699	0.107	0.101	0.261	−0.022	0.021	−0.009
Sum of squares	4.36	1.80	1.19	0.45	0.12	0.05	0.03	0.003
Cum. variance %	54	77	92	98	99	100	100	100

Table 6.8. Rotated factor loadings of the principal components. (Modified from Haan and Allen 1972) (Reprinted with permission from Water Resources Research, vol. 8, No. 9. Copyright 1972)

Variables	Principal Components							
	1	2	3	4	5	6	7	8
A	0.955	−0.162	−0.135	−0.199	0.021	−0.003	−0.042	0.042
S	−0.038	0.978	0.009	−0.137	−0.146	−0.011	−0.012	−0.000
L	0.782	−0.212	−0.465	−0.264	−0.018	0.238	−0.003	0.001
P	0.915	−0.100	−0.302	−0.164	−0.106	−0.017	0.150	−0.003
di	0.938	−0.157	0.262	−0.110	−0.068	−0.044	−0.077	−0.042
Rs	−0.094	0.050	0.988	0.110	0.002	−0.007	−0.006	0.010
F	−0.322	−0.196	0.175	0.909	−0.007	−0.011	−0.008	0.000
Rr	−0.407	0.839	0.103	−0.064	0.340	−0.011	−0.010	0.002
Sum of squares	3.52	1.81	1.41	1.01	0.15	0.06	0.03	0.004
Cum. variance %	44	67	84	97	99	99	100	100

6.11
Numerical Example on Sampling Sites

Multiple regression was applied in the determination of the most effective and cost-efficient sampling sites. Tasker and Stedinger (1989) extended generalized least squares (GLS) to the design of gaging networks and used the best method to address realities and complexities of regional hydrologic data sets. Simulation studies often do not treat these problems or methods. They mathematically treated and discussed the leverage and influence statistics for GLS regression so that one might use these statistics to identify sampling sites that extend an unusually large influence on a regression analysis. They defined the leverage of an observation and its influence on regression estimates to emphasize when new gaging sites are most influential and should be included in the gaging design network. Figures 6.3 and 6.4 show the results of the network analysis.

6.12
Multivariate Multiple Regression

Multivariate multiple regression is used when the objective is to regress a set of dependent variables X_1 on a set of independent variables X_2. Gilroy (1970) looked at the reliability of a variance estimate obtained from a sample augmented by multivariate regression. A sample of size N_1 of a normal random variable y is augmented by N_2 regression estimates obtained from p other random variables which along with y, follow a (p + 1) dimensional normal probaility law. He

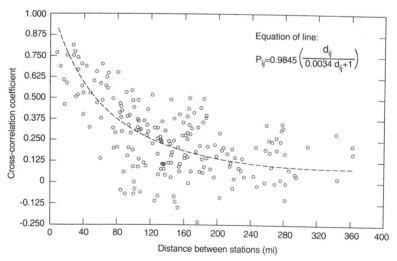

Fig. 6.3. Cross correlations of annual peaks for pairs of stations with at least 50 years of concurrent data in Illinois. (Tasker and Stedinger 1989) (Reprinted from Hydrology with kind permission from Elsevier Science-NL, Sara Burgerhartstraat 25, 1055 KV Amsterdam, The Netherlands. Copyright 1989)

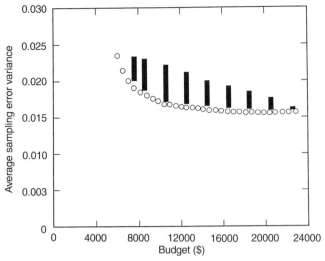

Fig. 6.4. Results of network analysis budget versus sampling error variance. (Tasker and Stedinger 1989) (Reprinted from Hydrology with kind permission from Elsevier Science-NL, Sara Burgerhartstraat 25, 1055 KV Amsterdam, The Netherlands. Copyright 1989)

derived the sampling variance of S_y^2, the estimate of the variance of y obtained from the augmented sample. He found that the relative reliability of S_y^2 (N_1), the variance based on the original N_1 observations, is dependent on N_1, N_2, p, and R, the multiple correlation between y and the remaining p variables. An application of this procedure was noted wherein the design of gaging stations was the objective. It was determined that given an existing set of gaging stations and faced with the decision of which subset of stations to retain, the network designer may wish to give consideration to estimating both the mean and the variance at discontinued sites (Gilroy 1970).

The treatment of this topic is not widespread in the literature and the reader is referred to Johnson and Wichern (1988) or other text for a fuller treatment and background. Only a few examples of application to geoscientific data were found but some applications to other data types are given in Johnson and Wichern (1988).

6.13
Nonlinear Regression

For many problems the linear model may be appropriate, at least as a first approximation to the true underlying model. For other problems a transformation to linearity might suffice for the parameters. There are many situations, however, when a linear model is not appropriate, for example, where the underlying model is a sum of exponential and/or trigonometric functions (Afifi and Azen 1972). In this case a transformation to a linear model is not easily done and an appropriate approximation cannot be made. In this case, an equation not convertible to linear form is said to be intrinsically nonlinear.

A model of this nature is termed a nonlinear regression model and can be written in the following form:

$$y_i = f(x_{1i}, \ldots x_{pi}; \quad \theta_1, \ldots, \theta_m) + e_i \qquad (6.6)$$

where $i = 1, \ldots, n$; and

$f(x_{1i}, \ldots, x_{pi})$ is a nonlinear function in the parameters $\theta_1, \ldots, \theta_m$, and e_i are uncorrelated error terms. When the underlying model is assumed to be linear, then the least squares estimators of the parameters are optimal since they are minimum variance unbiased estimators. However, when the model is nonlinear, i.e., nonlinear in the parameters, there are no such best estimators of the parameters. The method of maximum likelihood does, however, produce estimators θ_1 to θ_m which have desirable properties of being consistent and asymptotically efficient under some general conditions. Furthermore, if the error terms e_i are independent $N(0, \sigma^2)$, then these maximum likelihood estimators are also the least-squares estimators. In the case of a linear regression model, the least-squares estimators were obtained by solving a set of linear equations, but in the case of nonlinear models the equations are also nonlinear, and solution cannot be obtained in closed form. Because of linearity, it is necessary to use one of the iterative procedures to obtain the estimators (Afifi and Azen 1972). Also, the standard notation for nonlinear least squares situations is different from that for the linear least squares situation.

Draper and Smith (1981) discussed three methods for obtaining the parameter estimates. They are: (1) linearization (Taylor's series method), (2) steepest descent (sum of squares function method), and (3) Marquardt's compromise (uses both Taylor series and steepest descent features). Growth behavior is an example of a nonlinear model because it varies in time. The logistic model is a nonlinear model and takes several forms. For a more thorough treatment on estimating parameters, the reader is referred to Draper and Smith (1981). Box and Lucas (1959) also discussed the design of experiments involving the use of the linearization method for estimating parameters in the nonlinear case and this point is discussed in Draper and Smith (1981). Box and Draper (1965) and Erjavec et al. (1973) discussed situations in which parameter estimates are made for multiple responses and should be consulted for further discussion.

6.14
Multivariate Models with Qualitative (Dummy) Variables

6.14.1
Multivariate Models with One Qualitative Variable

During regression analysis, a simple linear model may be adequate. At other times, it may be often important to use the information from so-called qualitative variables to gain a better fit to the data. It is possible to estimate the correlation of continous variables (i. e., interval scale) with categorical (nominal scale) ones by using binary (or dummy variables) where classes are coded by using zero or one. Sometimes the dummy-variable method is useful when we want to index the relationship of a group or treatment variable with a dependent or differentiating variable. If more than two categories must be coded, each can be included as a separate dummy variable.

Sometimes, it is best to fit the data into homogeneous groups and fit two different models to the two or three groups, and this has the benefit of holding the qualitative variable constant for each group. It may be better however, in some instances, to use dummy variables and fit one model to all the data. The dummy variable is a variable that assumes only the value of zero or one. The dummy variable is used to indicate either the absence or presence of a particular qualitative characteristic of the observation. The use of the dummy variable adds a third term to the simple linear model and is determined as:

$$Y = \beta_0 + \beta_1 x + \beta_2 D \tag{6.7}$$

where $D = 0$ or 1.

When D has the value of 0, the third term is 0. When $D = 1$, the equation becomes:

$$Y = (\beta_0 + \beta_2) + \beta_1 x. \tag{6.8}$$

The result of this calculation is that only the intercept of the line is changed, since the slope is B_1. There is thus more flexibility in treating one dataset than treating separate groups. Sometimes to fit the data better, we use an interaction term in the equation, and if it involves a quantitative variable and dummy variable, such as Dx, the result is a change in slope of the regression line. For example, the equation

$$Y = \beta_0 + \beta_1 x + \beta_2 Dx \tag{6.9}$$

is the form of the equation.

When $D = 0$, the third term is 0 and drops out. When $D = 1$, the equation becomes:

$$Y = \beta_0 + (\beta_1 + \beta_2)x, \tag{6.10}$$

and the slope is changed. When the interaction term and the qualitative or dummy variable are used, then both the slope and intercept are changed. Thus the equivalent of two straight line fits can be obtained with the use of the general linear model approach with a dummy variable.

6.14.2
Multivariate Models with Multiple Qualitative Variables

The use of multiple dummy variables, D_1 and D_2, may be required, for instance, if instead of two types we have three or more. For example, when studying the price of ornamental stone, the following relation might be used:

$D_1 = 0, \quad D_2 = 0, \quad$ when mineral is quartz

$D_1 = 1$, $D_2 = 0$,   when mineral is feldspar
$D_1 = 0$, $D_2 = 1$,   when mineral is calcite.

This would lead to an equation of the form:

$$y = B_0 + B_1x + B_2D_1 + B_3D_2 + B_4D_1x + B_5D_2x ; \tag{6.11}$$

where this equation provides for a different slope and intercept for each type of mineral used.

Further, if for instance both mineral type and color are looked at, it would require two qualitative variables in the general linear model. This would require the addition of a third dummy variable, possibly, D_3, which could take on values of zero or one as well, for analysis. The general linear model equations become more complex as we add interaction terms for the two qualitative variables. Computer printouts give the parameter intercepts, estimated T for Ho: parameter = 0, $Pr > |T|$, and the standard error of estimate.

These models can be used with any set of data where the objective is to improve a fit to the data, because they can be fit with least squares calculations that have no underlying distributional assumptions. However, it is possible that terms will be included in a general linear model that do not make a significant improvement to the model. Formulating general linear models has been done using the method of least squares that has no underlying distributional assumptions. This method is to be contrasted with multiple linear regression where hypotheses about coefficients are tested, prediction methods are set up, and confidence intervals are determined. If the mean of a random variable Y is a function of other variables x's, the equation expressing the functional relationship : $E(Y) = f(x's)$ is a multiple regression equation, and if it is in the form of a general linear model as previously discussed, then it is a multiple linear equation. The general linear model for multiple regression includes an error term. In multiple linear regression, the tests are: (1) if individual coefficients = 0; (2) the overall model test (test all coefficients simultaneously) using the F-statistic; (3) a partial test of the model which allows several coefficients to be tested simultaneously; (4) what are the confidence intervals for B's terms, and (5) what are mean confidence intervals. The significance of variables may be done using a forward procedure, i.e., adding one significant variable at a time or backward elimination procedure, i.e., all variables are used then a systematic elimination is done. The backward elimination procedure may not be the best to use when the set of variables is large, i.e., if variables exceed a certain number of observations, and because, when a variable is dropped from the model, it does not have a chance to re-enter. We would alternatively use a procedure called stepwise regression, which finds the variable with the largest correlation with the dependent variable. A first model, a simple linear regression model, is thus set up with its R-squared value and probability. Probability values should be less than the specified value of 0.05 or the procedure will end. If $prob < 0.05$, then a two-variable model, three-variable model, and so on are set up by the program. Since backward elimination and stepwise regression procedures are different, they will not produce identical results.

6.15
Summary

This chapter has served to introduce in more detail multiple regression methods. Specific examples of the applications of these methods are summarized from the published literature.

6.16
Supplemental Reading

Afifi AA, Clark V (1990) Computer-aided multivariate analysis, 2nd edn. Van Nostrand Reinhold New York

Brown CE (1993) Use of principal-component, correlation, and stepwise multiple regression analyses to investigate selected physical and hydraulic properties of carbonate-rock aquifers. J Hydrol 147. 169–195

Draper NR, Smith H (1981) Applied Regression Analysis, 2nd edn. John Wiley, New York

Johnson RA, Wichern DW (1988) Applied multivariate statistical analysis. Prentice Hall, Englewood Cliffs, New Jersey

Tabachnick BG, Fidell LS (1989) Using multivariate statistics. Harper and Row, New York

CHAPTER 7

Multivariate Analysis of Variance

7.1
Concept

The multivariate analysis of variance (MANOVA) is a generalization of analysis of variance when there are multiple dependent variables, and tests whether mean differences among groups on a combination of dependent variables is a chance occurrence. A new dependent variable that maximizes group differences is created from the set of dependent variables by forming a linear combination of measured or observed dependent variables. Then an analysis of variance is performed on the new dependent variable by a test of variances.

7.2
Definitions

Analysis of Variance. Tests whether mean differences among groups on a single dependent variable are a chance occurrence.

Factorial Manova. A method wherein a different combination of dependent variables is formed for each main effect and interaction. Each combination then maximizes separation between treatment groups, gender groups, and interaction groups.

7.3
Overview of Analysis of Variance

7.3.1
Procedure

In this chapter, the consideration of procedures based on normal distribution theory will be extended to the analysis of data arising from designed experiments. We shall find that the calculation of a multivariate analysis of variance (MANOVA) is essentially similar to the calculation of a univariate analysis of variance. However, the testing of hypotheses and the interpretation of results is more complicated.

   A simple analysis of variance (AOV) is a method for testing the hypothesis that several different groups all have the same mean. In a designed experiment (one specifying order), observations are made following a plan which specifies the conditions under which each experimental unit is selected for measurement. The results are summarized in an analysis of variance table. A two-way analysis of variance is a test for two hypotheses. In some geoscientific studies, for example, the two hypotheses may be generalized or reduced to: (1) there is no significant difference between the rows (i.e., between strata in geology); and (2) there is no significant difference between columns (i.e., across strata in geology), or their alternatives.

For a multivariate analysis, p variates are measured on each experimental unit whether rock or water, giving a vector response, and not just the one (scalar) response as is the case for a univariate analysis. Note that the experimental plan or design specifies the selection of experimental units and so will be the same for each component of the response vector. The parameters of interest in AOV are: (1) treatment sum of squares, (2) blocks sum of squares, (3) error sum of squares, (4) mean squares for treatments and blocks, and (5) F-value. The cross product matrices must also be evaluated for MANOVA. All MANOVA tests are based on the matrix [H], the treatment sum of squares and cross-products matrix and [E], the error sum of squares and cross-products matrix. Some criterion values calculated for MANOVA are Roy's test criterion, Lawley and Hotelling's test, Pillai's test, and Wilks' likelihood ratio test (LRT) (see Johnson and Wichern 1988).

An analysis of variance between and within groups can be carried out in the usual way on each of the variates. In exactly the same way, the sum of products is defined as,

$$\sum (x_i\text{-mean}x_i)(x_j\text{-mean}x_j),\qquad\qquad\qquad\qquad (7.1)$$

and can be split up into components within and between groups. The analysis is as shown in Table 7.1.

The matrix [S] is expressed as the sum of matrices [B] and [W] of sums of squares and products between and within groups respectively, and is the basic form of the multivariate analysis of variance. It has exactly the same form as an ordinary analysis of variance and covariance, but the purpose is different in that the aim is not to carry out an analysis on one variate with the effect of the others removed, because all the x's are on the same footing. If a new random variable is defined as a linear function of the x's:

$$y = a_1x_1 + \ldots + a_px_p = \{a'x\}\qquad\qquad\qquad\qquad (7.2)$$

then it is easy to carry out an analysis of variance on y. The components are $\{a'Ba\}, \{a'Wa\}$, which sum to $\{a'Sa\}$.

Table 7.1. Form of analysis of variance

D. F.	Sums of products matrix	Var/Cov matrix
Between groups $(g-1)$	[B]	$[V] = [W]/(n-g+1)$
Within groups $(n-g+1)$	[W]	
Total n	[S]	

7.3.2
Repeated Measures Design in Analysis of Variance

A special case of the analysis of variance is called a repeated measures analysis of variance. The repeated measures design is used when the same variable is measured on several occasions for each subject or sample. The simplest repeated measures design is when two measurements are taken on each subject, i.e., pre- and post-test sample scores. The data are usually analyzed with a paired t-test. Besides requiring fewer experimental units, they provide a control on their differences, and variability due to differences between subjects can be eliminated from the experimental error (Norusis 1985). Multivariate tests of hypotheses and parameter estimates are also done during MANOVA procedures.

The analysis of variance includes two parts or structures that are important basic concepts. The basic concepts are design structure and treatment structure. The important basic design structures are divided into: (1) completely randomized, (2) randomized complete block, (3) latin square, (4) incomplete block designs, (5) split-plot, (6) repeated measures, and (7) crossover designs. The split-plot, repeated measures, and crossover design structures use the concept of different sizes of experimental units. The treatment structures are divided into (1) one-way, (2) two-way with controls, and (3) n-way structures. These methods are discussed in all statistical reference books and will not be discussed further in this text (see Griffiths 1967; Milliken and Johnson 1984).

7.3.3
Two-Way Multivariate Analysis of Variance Procedure

The two-way multivariate analysis of variance is similar to the univariate two-way analysis of variance. This method involves the use of at least two variables and two factor levels, with n-replications of the responses at each combination of factor levels (see Johnson and Wichern 1988).

7.3.4
Multiple Factor Analysis of Variance (Factorial Designs)

Multiple factor (multifactor) analysis of variance involves the study of more than one factor. What this means is that the design incorporates more than one independent variable, and each of these variables or factors can have two or more levels of its own. Usually we call such approaches 'factorial designs' and describe them in terms of the number of factors and the number of levels each has. For example a 2×2 design incorporates two factors each having two levels. A $4 \times 2 \times 2$ design has three factors, the first has four levels, the second factor has 2 levels, and the third factor has two levels. In a given design, factors and their levels define different subgroups in the experiment. Multifactor analysis of variance or factorial AOV provides methods for testing if different subgroups, or various combinations of subgroups, represent different populations in terms of what is being measured as the dependent variable. One of the useful features of a factorial AOV is that it allows us to test a number of different hypotheses in a single study.

If a study is done to evaluate the effects of two independent factors one at a time, for example, we call these effects "main effects". When we are asked to look at the effects of two independent variables in combination in a factorial desgn, we call these combinatorial effects "interactions". The nature of this interaction is the third hypothesis that can be tested in the present design. Most of the logic in calculating F ratios for a multiple-factor design is no different than for a univariate or single-factor model. Two-way and three-way analysis of variance have been applied to geologic populations by a number of scientists, and the reader is referred to Griffiths (1967) for a detailed and comprehensive treatment of these applications in the geological sciences.

The "latin-square" design or experiment is also described as one form of three-way or multifactor AOV which makes use of confounding to reduce the number of observations (Griffiths 1967). Factorial designs are handled by MANOVA and/or factorial discriminant programs in most software packages.

7.4
Overview of Manova

7.4.1
Canonical Variates Analysis

This is a technique for choosing those linear combinations of the treatment effect means which demonstrate the greatest inconsistency between the null hypothesis and the data. To demonstrate this inconsistency requires that we find coordinates of the sample means relative to axes of their subspace. The parameters to be evaluated are derived from MANOVA tables. Canonical variate plots of treatment means using the first two canonical variates may give an indication of the dimensionality of the data. Canonical variates analysis in two dimensions produces the best two-dimensional approximation in the least squares' sense.

7.5
Multivariate Analysis of Variance (MANOVA) Procedure

In a univariate analysis of variance, the total sum of squared deviations about the grand mean-written SS(Total) – is partitioned into a sum of squares due to one or more sources and a residual sum of squares. Associated with each partition is a number, called the degrees of freedom (d.f.), representing the number of linearly independent contrasts or alternatively the number of linearly independent parameters for that source. The partitioning is set out in an analysis-of-variance table.

In a p-dimensional multivariate analysis of variance based on the same design as the above AOV, there are p SS(Total)s to partition, one for each component, presented as sums of products. The MANOVA calculation is concerned with the partition of these measures of variance and covariance which are collected in a matrix of sums of squares and products written SSPM (Total). This is partitioned into sums of squares and product matrices due to the same sources as in the univariate case, and a residual sum of squares and products matrix. The matrices are all symmetric. Since the design of the experiment is the same, the degrees of freedom will be the same in the MANOVA as in the AOV.

The basic (noncomputer) method of calculation depends on the fact that the numbers in a particular position on the diagonals of the matrices are simply the sums of squares calculated for an AOV of the corresponding component of the vector response. For the off-diagonal terms, essentially the same formulae are used but squared terms are replaced by product terms for each pair of components. Four test statistics employed in some MANOVA tests are Pillai's trace, Wilks' lambda, Hotelling's trace, and Roy's largest root. The most 'robust' criterion in the presence of violations of variance and distribution assumptions is Pillai's trace (Norusis 1985).

7.6
Numerical Example on Water Chemistry

The multivariate analysis of variance technique was applied in the study of the chemistry of ground water in basaltic aquifers in Washington. Riley et al. (1990) conducted a statistical analysis of the hydrochemistry of ground-waters in the Columbia River basalts. They used multivariate cluster analysis, multivariate analysis of variance (MANOVA), canonical analysis, and discriminant analysis to investigate ground waters of the Saddle Mountains, Wanapum, and Grande Ronde formations near the Hanford Reservation in Washington. They found that the hydrochemistry of the ground waters in these basalts is distinctly different on each side of the Columbia River where the water passes through the Hanford Reservation site. MANOVA was performed on nine clusters to identify those variables that best distinguish the clusters and to assess the significance of cluster differences. Twelve variables were identified by the MANOVA procedure, with all clusters showing a significance after six steps. The first six steps identified magnesium, specific electrical conductance, manganese, silica, alkalinity, and potassium. Canonical rotation of 12 axes led to 3 canonical axes that account for 89% of the variabilities between clusters. A plot of the three canonical axes is depicted in Fig. 7.1 and 7.2. The symbols used on the figures are in accordance with Table 7.2. Figures 7.3 and 7.4 show the three dimensional clusters and spatial distribution characteristics of the data. Figure 7.5 shows the cluster membership versus cluster diameter for the data.

7.7
Numerical Example on Mine Tailings Hydrology

The multivariate analysis of variance was used to investigate the geochemistry of pore waters of mine tailings. In multivariate analysis of variance, we test that the means of two or more groups are significantly different. Using MANOVA as one of the multivariate procedures, Williams (1991) conducted a study of the geochemistry of pore waters in vadose (unsaturated) and saturated pore water zones in mine waste tailings. Williams (1991) used a variety of methods in the investi-

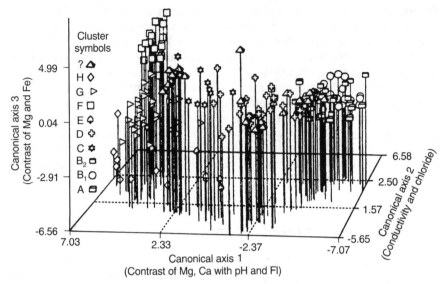

Fig. 7.1. Nine cluster canonical map for the combined off-site datasets. Symbols are given in Table 7.2. (Riley et al. 1990) (Reprinted from Hydrology, vol. 119, with kind permission from Elsevier Science-NL, Sara Burgerhartstraat 25, 1055 KV Amsterdam, The Netherlands. Copyright 1990)

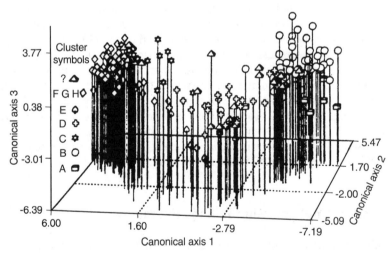

Fig. 7.2. Six cluster canonical map of the combined on-site and off-site datasets. Symbols are given in Table 7.2. (Riley et al. 1990) (Reprinted from Hydrology, vol. 119, with kind permission from Elsevier Science-NL, Sara Burgerhartstraat 25, 1055 KV Amsterdam, The Netherlands. Copyright 1990)

Table 7.2. Identification of clusters by predominant geologic formation and map symbols to accompany Figs 7.1 to 7.4. (Riley et al. 1990) (Reprinted from Hydrology, vol. 119, with kind permission from Elsevier Science-NL, Sara Burgerhartstraat 25, 1055 KV Amsterdam, The Netherlands. Copyright 1990)

Cluster symbol	Geologic Formation	Symbols used on canonical map		
		9 cluster map	6 cluster map	
A	Grande Ronde basalt	▣	▣	A
B1	Grande Ronde basalt	○	○	B
B2	Wanapum basalt	▣		
C	Saddle Mountain basalt	✳	✳	C
D	Deep Saddle Mountain basalt and Wanapum basalt	○	○	D
E	Grande Ronde basalt	◑	◑	E
F	Wanapum basalt and Grande Ronde basalt	□		F, G, H other
			◇ H	
G	Wanapum basalt	◁	▽	
H	Saddle Mountains basalt and Wanapum basalt ungrouped	◇ ▽		

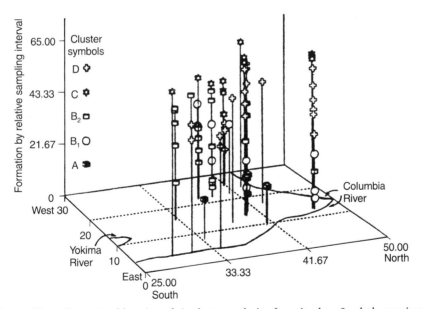

Fig. 7.3. Three-dimensional location of six cluster analysis of on-site data. Symbols are given in Table 7.2. (Riley et al., 1990) (Reprinted from Hydrology, vol. 119, with kind permission from Elsevier Science-NL, Sara Burgerhartstraat 25, 1055 KV Amsterdam, The Netherlands. Copyright 1990)

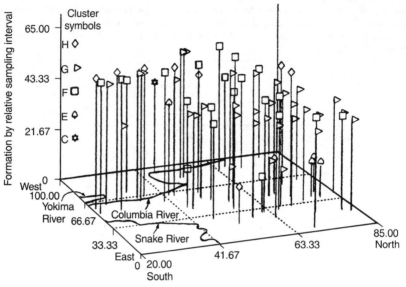

Fig. 7.4. Three-dimensional location of six cluster analysis of off-site data. Symbols are in Table 7.2. (Riley et al. 1990) (Reprinted from Hydrology, vol. 119, with kind permission from Elsevier Science-NL, Sara Burgerhartstraat 25, 1055 KV Amsterdam, The Netherlands. Copyright 1990)

Fig. 7.5. Cluster diagram for the combined on-site and off-site data sets using Q-mode hierarchical clustering. (Riley et al. 1990) (Reprinted from Hydrology, vol. 119, with kind permission from Elsevier Science-NL, Sara Burgerhartstraat 25, 1055 KV Amsterdam, The Netherlands. Copyright 1990)

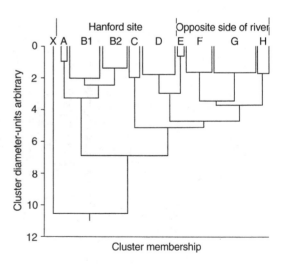

gations including principal components analysis, analysis of variance, and multivariate analysis of variance. We will briefly summarize findings from MANOVA. Several tests were run using MANOVA to study the two zones as independent variables to ascertain if the two zones exhibited significant differences. The most important variables determined from MANOVA are Pb, Zn, and S. Table 7.3 shows the results of principal components analysis on the data. Tables 7.4 and 7.5 give the results from univariate F-tests done in conjunction with MANOVA. The MANOVA results are given in Table 7.6. The two zones were found to be significantly different when Pb, Mn, Eh, Zn, Fe, and S were considered as leading variables in the tests.

7.8
Profile Analysis and Significance Tests

Profile analysis is defined here as a special case of T-squared tests or MANOVA. Profile analysis pertains to situations where measurements are done on two or more groups of samples. It is assumed that the responses for the different groups are independent of one another, but all responses must be expressed in similar units. Under MANOVA, we test whether the mean vectors are equal, but under

Table 7.3. Principal components in vadose and saturated zones. (Williams 1991)

Variable	Component Coeficients					
	1	2	3	1	2	3
	Vadose zone			Saturated Zone		
pH	−0.08	0.53	0.21	0.15	−0.36	0.08
Eh	0.03	−0.43	0.34	−0.23	0.34	0.06
Elements:						
Al	0.29	0.001	0.12	−0.10	0.49	0.06
B	0.28	−0.09	0.01	0.33	0.13	0.10
Ba	−0.005	0.04	0.61	0.07	−0.04	0.48
Ca	0.29	0.07	0.12	0.11	−0.24	−0.12
Cd	0.29	−0.04	−0.11	−0.12	0.47	−0.15
Cu	0.29	−0.01	0.05	−0.12	0.10	0.49
Fe	0.30	−0.08	0.04	0.34	0.15	0.07
K	−0.10	0.40	0.22	0.24	−0.07	−0.20
Mg	0.29	0.07	0.11	0.35	0.13	0.04
Mn	0.30	0.12	0.03	0.34	0.10	−0.03
Na	0.26	0.25	0.02	−0.04	0.01	−0.53
Ni	0.15	0.11	0.08	0.34	0.14	−0.001
Pb	0.16	−0.14	−0.46	0.06	0.20	−0.35
S	0.30	−0.02	0.06	0.33	0.16	0.06
Si	−0.08	−0.49	0.37	−0.08	0.25	0.08
Zn	0.29	−0.01	−0.05	0.35	0.10	0.01
Porportion of variability	0.59	0.12	0.08	0.44	0.16	0.10
Cumulative variability	0.59	0.71	0.79	0.44	0.60	0.70

Table 7.4. Univariate F-tests of zone and time interaction (hypothesis: zone-time interaction shows no significant effect). (Modified from Williams 1991).

Reject hypothesis (zone-time interaction is significant)			Cannot reject hypothesis (zone-time interaction not significant)		
Variable	F	Probability	Variable	F	Probability
Ba	2.66	0.0017	pH	1.56	0.089
Ca	1.92	0.0208	Eh	1.01	0.447
Pb	3.60	0.0001	Al	1.01	0.108
			B	1.45	0.134
			Cd	0.85	0.640
			Cu	0.62	0.585
			Fe	0.97	0.877
			K	1.02	0.440
			Mg	0.97	0.502
			Mn	0.82	0.668
			Na	1.41	0.139
			Ni	1.65	0.070
			S	0.67	0.836
			Si	0.88	0.589
			Zn	0.50	0.955

Table 7.5. Results of univariate F-tests of zone comparisons (hypothesis: means of each variable are the same in two zones). (Modified from Williams 1991)

Reject hypothesis (zones are significantly different)			Cannot reject hypothesis (zones are not significantly different)		
Variable	F	Probability	Variable	F	Probability
Eh	9.18	0.0191	pH	3.63	0.0986
B	5.91	0.0453	Al	0.10	0.7562
Fe	6.57	0.0374	Ba	0.91	0.3724
K	5.74	0.0478	Ca	1.01	0.3480
Mn	8.83	0.0208	Cd	0.97	0.3574
Ni	5.33	0.0542	Cu	0.11	0.7475
Pb	62.42	0.0001	Mg	4.63	0.0685
Zn	6.98	0.0333	Na	0.01	0.9269
			S	4.91	0.0622
			Si	4.91	0.0623

Table 7.6. Results of multivariate F-tests. (Williams 1991)

	F-statistic	Probability	Significant
Zone-time interaction	1.26	0.0858	No
Zone comparison	122.5	0.0081	Yes

Canonical coefficients: Eh = –3.8; Fe = –1.9; Mn = –0.8; S = 5.9; Pb = 9.4; and Zn = –7.8

Note: Other combinations considered but not presented here

profile analysis, we test the question of whether equality of mean vectors is divided into several specific possiblities (Johnson and Wichern 1988). Plots can be done to show the population profile. The hypotheses tested are: (1) parallelism – are the profiles parallel?; (2) coincidence – assuming that the profiles are parallel, are the profiles coincident?; and (3) levelness – assuming the profiles are coincident, are the profiles level, i.e., are the means equal to the same constant?.

The tests are illustrated in detail in Johnson and Wichern (1988) and clearly demonstrated in Tabachnick and Fidell (1989).

The test for parallelism may done on the scores (subtracted differences) and then a one-way analysis of variance can be done on the segments (slopes) to see if there is a slope difference between original dependent variables. If a slope difference is found, then the profiles are not parallel.

The test of flatness or levelness is done by looking at the combined groups, and whether the segments deviate from zero. We are then testing whether the segments (slopes) for the combined groups differ from zero (i.e., nonlevel or nonflat). The test of flatness is a multivariate generalization of the one-sample t-test and can be evaluated through a Hotelling T^2 test.

7.9
Profile Analysis – Discussion

Profile analysis is an application of multivariate analysis of variance (MANOVA) in which several dependent variables are measured on the same scale. It is also called the "multivariate approach to repeated measures" and is used in research studies where subjects are measured repeatedly (repeated measures) on the same dependent variables, but requires more measurements than its univariate alternative (i.e., the univariate repeated-measures ANOVA) (Tabachnick and Fidell 1989). The methods allow for testing of parallelism of profiles, overall difference among groups, flatness of profiles, and contrasts in profiles if more than two groups (or levels) are involved. In multiple time series designs, two groups – experimental and control – are tested on several pretreatment and several posttreatment occasions to provide multiple dependent variables. In these types of designs, profile analysis is also available as a substitute for univariate repeated measures ANOVA (Tabachnick and Fidell 1989). The requirement of measures on the same scale of measurement may be circumvented if standardized scores such as Z-scores are used instead of raw scores for the dependent variables. A complete example of the profile analysis procedure is found in Tabachnick and Fidell (1989) and the reader is referred there for a fuller treatment.

7.10
Comparison of Means and Simultaneous Tests on Several Linear Combinations

Means can be compared in many ways, through graphs, analysis of variance, T-square tests, and MANOVA tests. One such method of making inferences about a sample is an extension of a univariate confidence interval to a multivariate confidence region, and is best described in Johnson and Wichern (1988). Examples of simultaneous tests on several linear combinations can be found in Milliken and Johnson (1984). For this analysis to be valid, linear independence of the rows of the data matrix is assumed. The hypothesis, H_0, is tested using an F-statistic.

Many confidence intervals can be determined using the LSD multiple comparison method, Fisher's least significant difference (LSD) procedure, Bonferroni's method, multivariate t-method, Scheffe's procedure, Tukey's honest significant difference (HSD) procedure, Newman-Keuls method, Duncan's new multiple range method, and the Waller-Duncan procedure (see Milliken and Johnson 1984). A full discussion of this topic is extensive, and is well beyond the scope of this text, therefore the reader should consult Milliken and Johnson (1984) for test examples and comparison of the above procedures. The Bonferroni method is very useful for making a very small number of comparisons of means.

A confidence interval can be defined as an interval based on observations of a sample and so constructed that there is a specified probability that the interval contains the unknown true value of a parameter. For example, it is common to calculate confidence intervals that have a 95% chance of containing the true value. The confidence level is the degree of certainty or confidence associated with a confidence interval (i.e., the probability that the interval contains the true value of the parameter).

7.11
Calculating the Confidence Interval for the Difference of Two Means

7.11.1
Simultaneous Confidence Intervals

A situation may often arise that demands that a comparison of two populations is done with respect to some random variable. Let one sample, X_a, of size n_a have a mean, X_a, and a second sample, X_b, of size n_b have a mean, X_b. If X_a and X_b are normal random variables, then so are the mean: $\mu_a - \mu_b$ and variance: $Var_a/n_a + Var_b/n_b$. We define a new random variable as:

$$Z = \frac{(\bar{X}_a - \bar{X}_b) - (\mu_a - \mu_b)}{(\sigma_a^2/n_a + \sigma_b^2/n_b)^{1/2}},$$
(7.3)

which has a standard normal distribution. Then, using confidence intervals from standard normal distribution table:

$\Pr(-a < Z < a) =$ confidence interval.

Then we calculate confidence, c:

$$c = a\,[\sigma_a^2/n_a + \sigma_b^2/n_b)^{1/2}].$$
(7.4)

Continuing, we derive the following equation for confidence interval values:

$$(\bar{X}_a - \bar{X}_b) \pm c. \tag{7.5}$$

For large sample sizes, we can calculate variances, but for small sample sizes, we use the t-distribution.

Simultaneous confidence regions assess the joint knowledge about plausible values of mean u, and are statements about individual component means. The simultaneous T^2 statistic is the basis of simultaneous confidence statements which are related to the joint confidence region considered here.

7.11.2
The Bonferroni Method of Multiple Comparisons

An alternative method for multiple comparisons is the Bonferroni procedure (Johnson and Wichern 1988). The intervals developed using the Bonferroni procedure are often shorter under certain conditions.

7.12
Summary

This chapter has presented a summary of multivariate analysis of variance applications. The MANOVA method can be used in a number of different ways to test means, find confidence intervals, and to investigate profile characteristics.

7.13
Supplemental Reading

Afifi AA, Clark V (1990) Computer-aided multivariate analysis, 2nd edn. Van Nostrand Reinhold, New York

Johnson RA, Wichern DW (1988) Applied multivariate statistical analysis. Prentice Hall, Englewood Cliffs, New Jersey

Tabachnick BG, Fidell LS (1989) Using multivariate statistics, 2nd edn. Harper and Row, New York

CHAPTER 8

Multivariate Analysis of Covariance

8.1
Concept

The objective of multivariate analysis of covariance is to determine if there are statistically reliable mean differences that can be demonstrated among groups after adjusting the newly created variable (dependent variable) for differences on one or more covariates. When randomization assignment of samples or subjects to groups is not possible, multivariate analysis of covariance (MANCOVA) provides statistical matching of groups by adjusting dependent variables as if all subjects scored the same on the covariates.

8.2
Definitions

Covariates. Uncontrolled variables that are used to reduce variance due to error-from outside influences from error variance, thereby creating a more powerful test of mean differences among groups.

Multivariate analysis of covariance (MANCOVA). This is the multivariate extension of the analysis of covariance. MANCOVA addresses the hypothesis of whether there are statistically reliable mean differences among groups after adjusting the newly created dependent variable for differences on one or more covariates.

8.3
Overview of Analysis of Covariance

8.3.1
Procedure

The following short summary that discusses the major concepts of the analysis of covariance will serve to help introduce the subject of multivariate analysis of covariance.

The one-way analysis of covariance utilizes both the concepts from analysis of variance and simple linear regression. For instance, a measured factor, A, (also called a treatment factor) may have several levels (I). If we denote the y_{ij} as the observation made on the jth experimental unit for the ith level of A, $j = 1, \ldots, J_i$, $i = 1, \ldots, I$, and if y_{ij} is assumed to be normally distributed ($N(\mu_i, var)$, then we derived the one-way analysis of variance model (Afifi and Azen 1972) where:

$$y_{ij} = \mu + \alpha_i + e_{ij}, \tag{8.1}$$

with $j = 1, \ldots, J_i; i = 1, \ldots, I; \mu$ = overall or grand mean; α_i is the differential effect of the ith level of factor A; $\mu_i = \mu + \alpha_i$; and e_{ij} are independent $N(0, \sigma^2)$ error variables.

We now impose further that:

$$\sum_{i=I}^{j} J_i \alpha_i = 0 \tag{8.2}$$

in order to obtain unique least-squares estimators for μ and $\alpha_1, \ldots, \alpha_I$. If we also measured another variable, called a covariate or concomitant variable (i.e., linearly related to y_{ij}, for the jth experimental unit at level i of the factor, then we would write the model as:

$$y_{ij} = \mu + \alpha_i + \beta(x_{ij} - \bar{x}..) + e_{ij}, \tag{8.3}$$

with $j = 1, \ldots, J_i$, $i = 1, \ldots, I$

and where:

$$\bar{x}.. = \frac{1}{n} \sum_{i=1}^{I} \sum_{j=1}^{ji} x_{ij},$$

with $n = \sum_{i=1}^{1} J_i.$ \hfill (8.4)

This then becomes the model for the one-way analysis of covariance and expresses the ijth observation as a sum of the overall mean μ, a fixed differential effect α_i due to the ith level of the factor, a term $\beta(x_{ij} - meanx..)$ attributable to the linear association with x_{ij}, and an error term e_{ij}. If written as a one way analysis of variance equation, we derive:

$$y^* = \mu + \alpha_i + e_{ij}, j = 1, \ldots, J_i, i = 1, \ldots, I, \tag{8.5}$$

where:

$$y^*_{ij} = y_{ij} - \beta(x_{ij} - \bar{x}..) \tag{8.6}$$

which is the y_{ij} after adjustment for the linear regression on x_{ij}. Thus α_i may be regarded as the true differential effect of the ith level of factor A after adjustment for the linear regression on the covariate x_{ij} (Afifi and Azen 1972). There are many advantages of this model, such as increasing precision of the measurement of interest y_{ij} by taking out the effect of a related pretreatment variable x_{ij}. The gain in precision depends on the size of the correlation between the two variables. An example of a computer application with output and calculations of a model setup may be found in Afifi and Azen (1972) and will not be presented here.

8.3.2
One-Way Model with Multiple Covariates

The above analysis for the one-way analysis of covariance can be extended to a one-way model with multiple covariates. The equation is then written as:

$$y_{ij} = \mu + \alpha_i + \beta(x_{ij} - \bar{x}..) + \gamma(z_{ij} - \bar{x}..) + ... + e_{ij}, \tag{8.7}$$

where $j = 1, ..., J_i$, and $i = 1, ..., I$.

Here $x_{ij}, z_{ij}, ...$, are covariates, each linearly related to y_{ij}. The analysis would estimate μ, each of the coefficients $\beta, \gamma, ...$, and the adjusted differential effects α_i. We could then test whether these effects are equal, using the model.

8.3.3
M-Way Model with Covariates

The one-way analysis of covariance can be extended to an m-way model with covariates. The model equation is of the form (Afifi and Azen 1972):

$$y_{ij} = \mu + \alpha_i + \beta_j + \gamma(x_{ij} - \bar{x}..) + e_{ij}, \tag{8.8}$$

which has two factors and one covariate linearly related to y_{ij}. The reader is again referred to Afifi and Azen (1972), and other works (Scheffe 1958) and (Cochran 1957) for additional insight and relevant discussions.

Computer output can result in tests of pretreatment means, posttreatment means, adjusted means, and a test of $\beta = 0$. This total analysis includes the following:

1. Tables of the sum of squares and cross products for the sources of variation between levels (group means), within levels (subgroup means – which are also called the residual or error), and the total;
2. Estimates of means and parameters;
3. For each level i, a within group regression line, which is a least squares line for the subpopulation corresponding to the ith level of the factor;
4. Mean regression lines, of which the first is the least squares line through the sample means, and the means regression coefficient equal to the ratio of between means sum of squares for xy and xx, and the second is the total regression line, which is the regression of y on x when the samples for each of the I levels are combined into one sample of size n, and the total regression coefficient equal to the ratio of the total sum of squares for xy and xx;
5. An unbiased estimate of the error variance;
6. Calculated estimated adjusted differential effects of each level, i.e., α_1 to α_I, the differential effect of the first treatment through all adjusted by the regression of y on x; and
7. Estimated within group regression equations.

The tests of hypotheses may include:

1. That the I population means of the covariate are all equal (by appropiate F-test). This is a test of randomness of the assignment of experimental units to the I levels of the factor;

2. That the I population means of the observation variable are all equal (also by F-test);
3. That the I population means of the adjusted variable are all equal (by F-test in partitioned ANCOVA table); and
4. That $\beta = 0$, i.e., that the within group coefficient is zero (which is the ratio of SS due to regression/deviations about regression mean square error, MSE) (Afifi and Azen 1972).

8.4
Special Tests on Covariance Matrices

The analysis of covariance (ANCOVA) is commonly done as an analysis of variance and regression analysis combined. For the analysis of covariance to be an improvement over the analysis of variance, the common slope (B) must not $= 0$, and we test the slope with an F-test before proceeding to ANCOVA. The analysis of covariance is often used as a procedure to adjust the analysis for variables that could not be controlled by the experimenter. In this text, the analysis of covariance may also be used as a procedure for comparing several regression lines or surfaces, one for each treatment or treatment combination, where there is possibly a different regression surface for each treatment or treatment combination. There are many research situations in which one covariate is measured on each experimental unit, giving rise to the model equation for each treatment as (Milliken and Johnson 1989):

$$y_{ij} = a_i + \beta_{1i}x_{1ij} + \beta_{2i}x_{2ij} + \dots + B_{ki}x_{kij} + e_{ij}, \tag{8.9}$$

where x_{pij} denotes the value of the pth covariate on the $_{ij}$th experimental unit; a_i denotes the intercept of the ith treatment's regression surface; B_{pi} denotes the slope in the direction of the pth covariate for treatment i; $e_{ij}, j = 1, 2, \dots, n_i$, $i = 1, 2, \dots, t$ are independent and identically distributed normal with $N(0, \sigma^2)$.

The previous equation is that of a k-dimensional plane in $k + 1$ dimensions which is assumed to adequately describe the data for each treatment. The usual regression diagnostics should be applied to each treatment's data to check for model adequacy. The treatment variances should be tested for equality. When there is only one covariate, the analysis of covariance consist of comparing several regression lines, and for multiple covariates, the analysis of covariance consists of comparing several regression planes or hyperplanes. Many forms of models just like multiple regression are possible, such as a polynomial function of one independent variable, and we would be comparing polynomial regression models from several treatments. Further, the models could be a quadratic function of two variables (or a quadratic response surface) as we would be comparing t response surfaces (Milliken and Johnson 1989).

Tests of various hypotheses may involve the following:

1. tests that all slopes for hth covariate are zero;
2. tests that all slopes for the hth covariate are equal, meaning that the surfaces are parallel in the direction of the hth covariate.

For additional treatment of these topics, the reader is referred to Milliken and Johnson (1989). Many examples in agricultural studies are also available on this subject. An analysis of covariance applies specifically if both qualitative and quantitative supplementary data are available, and the objective is to see if one of the qualitative variables is a significant contributor to the model. The extensions of these methods to the study of several independent variables is straightforward. If the ANCOVA is justified and leads to a significant F-test for differences among adjusted means, then we follow with a test that compares the means, a multiple comparison procedure.

8.5
Multivariate Analysis of Covariance Procedure

The multivariate analysis of covariance is an extension of the analysis of covariance with multiple covariates. As with other methods such as factor analysis and MANOVA, MANCOVA has hypothesis testing as a part of the analysis. The MANCOVA procedure is concerned with the two residual matrices. Davis (1986) illustrated the use of variance-covariance tests for equality of matrices. To compute the test statistic, the determinant, and the matrices of variances and pooled variances are determined. For example, if measurements for k populations of observations are made, a variance-covariance [var-cov] matrix may be computed and a test of generalized variances or multivariate F test can be done.

The test is: Ho; $[E_1^2] = [E_2^2] = \ldots = [E_k^2]$, i.e., that all covariance matrices are equal; and alternative hypothesis: Ha; $[E_i^2]$ does not equal $[E_j^2]$, i.e., at least two are different. If we take the k populations to be equal then the sample variance-covariance matrix is an estimate of the the population variance-covariance, and we can determine a pooled estimate of the variance-covariance matrix according to:

$$s_p^2 = \sum_{i=1}^{k} \frac{(n_i - 1)\,[s_i^2]}{(\sum_{i=1}^{k} n_i) - k}, \tag{8.10}$$

where n_i is the number of samples in the ith group and SUMn$_i$ indicates the grand total of all samples in all k groups.

From the pooled estimate of the variance-covariance matrix, a test statistic M is computed, where:

$$M = \left[\left(\sum_{i=1}^{k} n_i\right) - k\right] \ln|s_p^2| - \sum_{i=I}^{n}\left[\left(n_i - 1\right)|n|s_i^2|\right] \tag{8.11}$$

The test is based on the logarithm of the determinant of the pooled variance-covariance matrix and the average of the logarithms of the determinants of the sample variance-covariance matrices. M is converted to a chi-square statistic and the approximation is good if k and m do not exceed about 5 and each variance-covariance estimate is based on at least 20 observations. The transformation is derived as (Davis 1986):

$$C^{-1} = 1 - \frac{2m^2 + 3m - 1}{6(m+1)(k-1)} \left(\sum_{i=1}^{k} \frac{1}{n_i - 1} - \frac{1}{\sum_{i=1}^{k} n_i - k} \right) \tag{8.12}$$

where:

chi-square statistic $= MC^{-1)}$. $\hspace{5cm}$ (8.13)

The approximate chi-squared value has degrees of freedom equal to:

$$v = (1/2) \ (k-1)m(m+1). \tag{8.14}$$

If groups are equal then (8:12) becomes:

$$C^{-1} = 1 - \frac{(2m^2 + 3m - 1)(k+1)}{6(m+1) \ k(n-1)}, \tag{8.15}$$

where k = populations, and C^{-1} is compared with critical chi-square value based on (8.14).

Davis (1986) gives an excellent example of the application of analysis of variance-covariance matrices for two aquifer systems composed of limestone and alluvium in Kansas. The reader is referred there for more detail.

8.6
Numerical Example – Tests on Covariance Matrices

Two covariance matrices can be tested for equality in other ways. Muirhead and Waternaux (1980) described one test; that a covariance matrix (A) equals a given covariance matrix (B). They used a likelihood ratio test, Λ, of the form (Anderson 1958):

$$\Lambda = \frac{|S_1|^{(1/2N_1)} |S_2|^{(1/2N_2)}}{|S_1|^{1/2N}}, \tag{8.16}$$

where S_1 and S_2 are two sample covariance matrices, and $||$ is determinant (see also Agresti 1990).

Many other tests on covariance matrices can be found in the literature. Among these are tests of structure of matrices (i.e., [COV] = fixed positive definite symmetric matrix) and if equality exists (Muirhead and Waternaux 1980). Muirhead and Waternaux (1980) discussed the testing of the structure of covariance matrices using a likelihood ratio statistic as given by Anderson (1958). Johnson (1987) used a modified LRT to test the equality of several covariance matrices with k independent samples. Other numerical examples of tests include covariance matrices tests for block diagonal property using a likelihood ratio test (LRT) (Muirhead and Waternaux 1980). Johnson (1987) used an LRT, V_1, to do a generalized test of independence (whether we have a diagonal matrix) of the form:

$$V_1 = \frac{|Cov|}{\prod\limits_{i=1}^{P} Vqr_{ii}} = |R| \tag{8.17}$$

The hypothesis in general consisted of a test: [Cov] = diagonal matrix, versus Ha: that Cor_{ij} not equal to 0, for some i not equal j. The LRT, V_1, is then related to critical points. Equation 8.18 describes another covariance matrix test of independence as given in Johnson (1987) and is the derivation of a more generalized test of independence (i.e., whether submatrices of covariance matrix equals 0) using:

$$V_2 = \frac{|R|}{\left[\prod\limits_{i=1}^{k} |R_{ii}|\right]} \tag{8.18}$$

with R_{ii} being submatrices of R matrix. Critical points are in Korin (1968) and Nagarsenker and Pillai (1973).

Numerical examples of covariance matrix tests of sphericity (i.e., COV $= \lambda [I]$) also exist and a simple example of a sphericity test would be a biological test of whether residue in the body is independent of location in the body for some chemical residue being measured. A test for sphericity uses the Ho: Cov = Var[I] versus Ha: Cov does not equal Var[I], where [I] = identity matrix. The test is structured as:

$$V_3 = \frac{|W|}{(\frac{1}{p} tr\, W)^p} \tag{8.19}$$

where W = (n−1) Cov, and trW = trace W (Johnson 1987) (see also Agresti 1990).

A more complete analysis of tests on covariance matrices can be found in Muirhead and Waternaux (1980) and the reader is referred there.

8.7
Summary

In this chapter, a brief description has been given of the various MANCOVA methods. Many different forms of likelihood ratio tests are used for testing covariance matrices for structure and equality.

8.8
Supplemental Reading

Davis JC (1973) Statistics and data analysis in geology. John Wiley, New York
Tabachnick BG, Fidell LS (1989) Using multivariate statistics. Harper and Row, New York
Afifi AA, Azen SP (1972) Statistical analysis – a computer oriented approach. Academic Press, New York

Variable-Directed Techniques not Based on Normal Distribution Assumptions

CHAPTER 9

Principal Components

9.1
Concept

Variables that are correlated to one another form factors or principal components related to an underlying or theoretical process.

9.2
Definitions

The definitions in Chapter 4 are the same definitions used in this section (see Sect. 4.2).

9.3
Overview and Objectives

The objective of principal components analysis (PCA) is to determine the relations existing between measured properties that were originally considered to be independent sources of information. Principal components are the eigenvectors of the variance-covariance matrix, developed from the original data matrix. Through evaluation of the principal components, one seeks to determine the minimum number of variables that contain the maximum amount of information and to determine which variables are strongly interrelated. The physical significance of the interrelations of the components of the data are sought to provide a simple interpretation of processes causing variation in variables. This technique was developed by Hotelling (1933) from original work of Pearson (1901). The main parameters of concern in PCA are factor loadings, and the sum of explained variation.

The main concern is if there are theoretical or underlying variables that relate to processes being studied? Can redundancy in the data be reduced?

Several papers that have been discussed have also indirectly focused on some of the procedures covered in this part. The procedures addressed here in general are principal components. Other papers that treat the methodology inherent in principal components and discussed earlier in this volume are Griffiths (1966), Rauch and White (1970), Haan and Allen (1972) and McCuen et al. (1979). Therefore, only a few papers are discussed for this subject (Hull 1984; Brown 1993).

9.4
Principal Components Analysis Procedure

Principal components analysis has been found to be important in understanding more fully the processes that control the formation of rocks and the contributing elements to the geochemistry of rocks and water. Mathematically, it is also the starting point for most factor analysis models. It is described in the literature either as "principal component" analysis or "principal components" analysis, whichever is more preferable to the reader.

In any study using principal components analysis (PCA), the data consist of $n = x$ observations on $p = y$ variables. From this n x p matrix, we compute a p x p matrix of correlations. In essence, principal components analysis extracts p roots or eigenvalues, and p eigenvectors from the correlation matrix. The number of roots corresponds to the rank of the matrix, which equals the number of linearly independent vectors. The eigenvalues are numerically equal to the sums of the squared factor loadings and represent the relative proportion of the total variance accounted for by each component (Griffiths 1966; Davis 1973). Principal component factor loadings can be rotated to enhance interpretation, but ease of interpretation is not always the result of rotation (see Davis 1973). In an analysis, rotation of the loadings is systematically done but includes only the number of significant principal components determined using the size of eigenvalues (equal to or greater than one) or the percent of explained variation (less than or equal to 95%). Rotated as well as unrotated loadings are considered in most papers for clarity. Given p variables and n samples, methods are termed R-mode when interrelations between variables are the concern, and Q-mode when the concern is interrelations between samples.

Normally distributed and non-normally distributed data can be subjected to principal components analysis as opposed to factor analysis, which assumes a statistical model based on normal-distribution theory, whereas PCA does not. The principal components analysis program used in most studies is similiar to the program developed by Ondrick and Srivastava (1970). The correlation matrix of original data is factored by maximizing the variance around independent axes as done by Hotelling (1933). The principal components are orthogonal and sequentially ordered in terms of the amount of variance that each contains, such that the first component has the largest amount of variance. The smaller components may not contain significant portions of the total variance contained in the data and are excluded from further use in the study.

9.5
Defining Sedimentary Processes

Because a sedimentary rock is the end result of a set of dynamic sedimentary processes such as weathering, erosion, transportation, deposition, lithification, and cementation, the interaction of those processes accounts for the resulting properties of the rock and may be appropiately called a "factoral" process. Griffiths (1966) stated that in PCA those variables that load heavily on a given component or factor can be assumed to be the result of the same process or processes, inasmuch as all of the variables that correlate and load on the same component are measures of the same "factoral" processes. The principal components may be physically interpretable in terms of those variables that have high factor loadings on them. A component is thus physically interpretable only within the context of the variables that are either strongly or weakly associated with that particular component.

The key parameters to be analysed in PCA are sum of variation explained, eigenvalues greater than one, and cumulative variation explained, because these parameters denote when factors or components are nonsignificant in the statistical sense and should be discarded. The cumulative variation explained is expected to be in the 95% range, and the eigenvalues should have values greater than one, otherwise the factors are probably the result of statistical noise and more data is needed.

9.6
Numerical Example on Ground Water Geochemistry

A principal components analysis was used to study the chemistry of ground-water in the Sacramento Valley, California. Hull (1984) studied the geo-chemistry of ground-water in the Sacramento Valley, California using data on 15 geochemical variables – dissolved solids, calcium, potassium, magnesium, sodium, arsenic, boron, silica, bicarbonate, sulfate, chloride, nitrate, phosphate, and fluoride – that were collected and subjected to principal components analysis. Table 9.1 provides the results of this study. The results indicate that only two principal components are important in describing the variance in the chemical data.

Component 1 is an underlying variable (factor) that denotes source area, because it has loadings from all the major chemical variables. Component 2 is an underlying variable (factor) that relates the occurrence of fine-grained sediments and reducing conditions and is an important control on water chemistry in the aquifer. Two components account for over 50% of the variation in chemical variables measured in the Sacramento Valley. In the central part of the valley, flood-plain deposits are finer grained and under reducing conditions, whereas sediments along the valley walls are coarse grained and oxidized (Hull 1984). Much of the variance in the data set is not explained by two principal components.

9.7
Numerical Example on Properties of Carbonate Rock Aquifers

A principal components analysis was completed on geohydrologic data collected from carbonate rock aquifers in central Pennsylvania. Brown (1993) synergistically used several multivariate procedures to evaluate the importance of laboratory-measured variables and to assess the aquifer potential at the intergranular scale of selected carbonate rocks. Principalcomponents analysis was used to

Table 9.1. Results of principal components analysis on 15 chemical variables from the Sacramento Valley, California. (After Hull 1984)

	Principal Components		Communalities
	1	2	
Dissolved solids	0.94		0.3
Calcium	0.76		0.60
Magnesium	0.81		0.67
Sodium	0.83		0.76
Bicarbonate	0.85		0.72
Chloride	0.76		0.62
Sulfate	0.63		0.41
Boron	0.65		0.42
Fluoride	0.30	−0.46	0.30
Silica	−0.54	0.42	0.46
Potassium		0.67	0.48
Nitrate		−0.71	0.52
Phosphate		0.50	0.30
Manganese		0.63	0.41
Arsenic		0.62	0.38
Percent Variation	36	17	
Cumulative Variation (%)	36	53	

reduce the number of significant variables while also identifying and showing that some variables were related to the same hydrologic processes. Correlation analysis in conjunction with principal component and regression analyses was used to study the aquifer behavior of selected carbonate rocks. Four principal components were found to be significant in explaining the variance of the data. Stepwise multiple regression analysis was used to predict porosity and permeability from measured variables and the regression was found to be significant at the 5% significance level. Tables 9.2 and 9.3 show the unrotated and rotated factor loadings and the sum of variation explained (in percents). Between 90 and 95% is taken as the cutoff point for denoting significance of a factor. This cutoff level denotes that four or five factors are significant, and the rest are probably meaningless. Figure 9.1 is a plot of the variation explained to the principal components; the slope begins to flatter significantly between four and five, denoting a significant change in variation explained.

9.8
Uses of PCA

9.8.1
Generalizations

PCA is appropiate only when all the variables arise on an equal footing. If the response variables are measured in the same units, then changing the scale of measurements on one or more of the variables will have an affect on the principal components of possibly reversing the role of important versus un-

Tables 9.2. Principal components matrix – unrotated for all carbonate formations. Values larger or smaller than ± 0.355 are significant. (Brown 1977; 1993) (Reprinted from Hydrology, with kind permission of Elsevier Science – NL, Sara Burgerhartstraat 25, 1055 KV Amsterdam, the Netherlands, Copyright 1993)

Variables[a]	Factors								Comm.
	F1	F2	F3	F4	F5	F6	F7	F8	
1	-0.527	-0.028	0.761	0.254	-0.552	-0.127	0.239	-0.018	1.0
2	-0.544	0.427	-0.493	0.183	-0.344	0.339	0.108	-0.006	1.0
3	0.023	0.794	-0.250	0.170	-0.376	-0.366	-0.051	-0.004	1.0
4	-0.682	-0.357	-0.509	0.312	-0.191	-0.112	0.022	-0.006	0.9
5	0.682	0.357	0.508	-0.312	-0.191	0.112	-0.024	0.006	0.9
6	0.103	-0.932	0.056	0.140	-0.307	-0.031	-0.041	0.003	0.9
7	0.627	0.175	0.073	0.739	0.078	0.050	0.010	0.125	1.0
8	0.131	-0.937	0.022	0.119	-0.298	0.003	-0.027	-0.018	0.9
9	0.695	0.113	0.085	0.685	0.069	0.071	-0.049	-0.127	1.0
10	0.06	-0.126	-0.317	-0.199	0.004	-0.041	0.144	-0.012	0.9
11	-0.843	0.170	0.453	0.158	-0.029	0.082	-0.146	0.001	0.9
Sum of squares	3.94	2.90	1.73	1.44	0.531	0.308	0.120	0.034	
Variation explained %	35.8	26.4	15.7	13.0	4.83	2.80	1.09	0.304	
Cumulative %	35.8	62.22	77.90	90.95	95.77	98.57	99.66	99.96	

Comm., Communality

[a] Variables:

1 Bulk density, (g/cc);
2 Porosity, %;
3 Log permeability, in millidarcies;
4 Insoluble residue, %;
5 Total carbonate, %;
6 Grain length of long a-axis, φ units;

7 Standard deviation of a-axis length;
8 Grain length of short b-axis, φ units;
9 Standard deviation of b-axis length;
10 Calcite, %;
11 Dolomite, %.

Table 9.3. Rotated matrix of factor loadings. Values larger than ± 0.355 are significant. (Brown 1977, 1993) (Reprinted from Hydrology, with kind permission of Elsevier Science – NL, Copyright 1993)

Variables[a]	Factor Loadings						
	F1	F2	F3	F4	F5	F6	Comm.
1	-0.937	-0.088	0.049	-0.001	0.133	0.027	0.07
2	-0.120	0.263	-0.327	-0.115	-0.861	-0.234	0.99
3	0.036	0.450	0.074	0.140	-0.242	-0.844	0.99
4	-0.135	-0.130	-0.960	-0.149	-0.139	0.035	0.99
5	0.135	0.129	0.60	0.148	0.139	-0.035	0.99
6	0.044	-0.971	-0.121	0.018	0.124	0.150	0.98
7	0.125	0.019	0.121	0.73	0.038	-0.074	0.84
8	0.093	-0.966	-0.116	0.015	0.102	0.179	0.98
9	0.179	-0.038	0.173	0.55	0.072	-0.040	0.82
10	0.880	-0.134	0.317	0.226	0.174	0.014	0.74
11	-0.927	0.146	-0.160	-0.210	-0.155	0.014	0.75
Sum of squares	2.62	2.23	2.16	2.03	0.4	0.83	
Variation explained	23.9	20.3	19.6	18.5	8.6	7.6	
Cumulative %	23.9	44.1	63.7	82.2	90.8	98.3	

[a] Variables:

1 Bulk density, g/cc;
2 Porosity, %;
3 Log permeability, in millidarcies;
4 Insoluble residue, %;
5 Total carbonate, %;
6 Grain Length of a-axis, in φ units;
7 Standard deviation of a;
8 Grain length of b, φ units;
9 Standard deviation of b;
10 Calcite, %;
11 Dolomite, %.

Figure 9.1. Graph showing amount of variation explained by principal components. (Brown 1977)

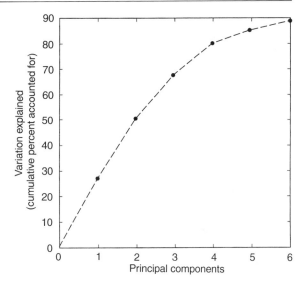

important variables. PCA should not be used when the analysis calls for some dependent variables and several explanatory or independent variables without thorough consideration of results. There is often a strong tendency to give meaning or to name the components, but this should only be done if the interpretations are obvious to the investigator. More often than not the interpretations are not obvious, so the naming process is done with caution. PCA should generally not be used to eliminate variables as the primary goal. PCA transforms a set of correlated variables to a new set of uncorrelated variables. If the correlation matrices seem to indicate little or no significant correlations, then PCA should not be done. PCA assumes no probability distribution on the data, although more meaning can be given to the components when the original data are multivariate normal. In the case where there are only two or three main components, the scores of these components for each individual can be plotted. This may indicate outliers or clusters in the data. Principal components are generally changed by scaling, thus they are not a unique characteristic of the data. If one variable has a much larger variance than all other variables, then it will tend to dominate the first principal component of the covariance matrix regardless of the correlation structure. There is little point in carrying out a PCA on the "covariance matrix" unless the variables have roughly similar variances, and the PCA on the correlation matrix is a better choice. The principal components are the eigenvectors of the correlation matrix, and are not the same as the eigenvectors of the covariance matrix. Choosing to analyze the correlation matrix rather than the covariance matrix involves deciding that all response variables are equally important, so the decision should be made carefully. By default, most statistical computing packages analyze the correlation matrix rather than the covariance matrix. Principal component scores are calculated differently when the PCA is done on the the correlation matrix rather than the covariance matrix.

How many principal components are important may be determined by several methods. The first way is to choose a significance of 95% as the significant amount of total variability that is being sought. The second way is to use a plot of the eigenvalue on one axis versus the number of the eigenvalue on the other axis, giving rise to a "scree plot". This second method determines a curve that flattens out, indicating a significant loss of variability per eigenvalue. The third method is used only if the PCA is performed on a correlation matrix. Its aim is to use as significant only the eigenvalues larger than one. If some of the original variables are linearly dependent, then some eigenvalues will be zero. One drawback to PCA when the data is non-normal is that there is no objective way to decide how many components to retain, so normality of data is a good characteristic of data for PCA as well.

9.8.2
Other Uses of PCA

Many uses of PCA are discussed in the literature. These uses will be generalized below.

9.8.2.1
Deriving Discriminant Functions

PCA can be used to determine discriminant functions in cases where the data matrix is singular and inversion of the matrix is a problem causing one to improvise, such as when there may be many more variables than observations.

9.8.2.2
For Multiple Regression

Multiple regression can be dangerous when the so-called independent variables are highly correlated (collinearity). One might consider regressing the dependent variable on the first few important principal components rather than on the original variables.

9.8.2.3
For Analysis of Variance

In experimental design type problems, one might consider doing a univariate analysis of variance on the component scores rather than univariate analysis of variance on the original variables. This must be done on the covariance matrix however, because in 2-d space, we are comparing two populations.

9.9
Summary

This chapter has defined some important uses of principal components analysis which is an important method for looking at large datasets and reducing dimensionality of the data (the number of important variables). Other examples with geological or hydrological data can be found elsewhere.

9.10
Supplemental Reading

Brown CE (1993) Use of principal components, correlation, and stepwise multiple regression analyses to investigate selected physical and hydraulic properties of carbonate-rock aquifers. J Hydrol 147: 169–195

Davis JC (1986) Statistics and data analysis in geology, 2nd edn. John Wiley, New York.

Chatfield C, Collins AJ (1980) Introduction to multivariate analysis. Chapman and Hall, New York

Individual-Directed Techniques Based on Normal Distribution Assumptions

CHAPTER 10

Multiple Discriminant Analysis

10.1
Concept

The objective of discriminant analysis is to determine group membership of samples from a group of predictors by finding linear combinations of the variables which maximize the differences between the populations being studied, with the objective of establishing a model to sort objects into their appropriate populations with minimal error.

10.2
Definitions

Discriminant Functions. A linear combination of weighting coefficients and standardized values of discriminating variables. It is one less than the number of groups being compared.

Centroid. Mean discriminant scores for each group on each function.

Canonical Correlation. Correlation between a discriminant function and the groups.

Wilks' Lambda. A calculation used to determine if amounts of variance accounted for by discriminant variables are significant. A problem that arises quite often in science is to discriminate between two groups of individuals or objects on the basis of several properties of those individuals or samples.

10.3
Overview

In geohydrology, for example, a hydrologist may want to classify a water sample into one of two classes based on measured chemical properties. When two or more variables are used to predict membership in categories or groups, the method is known as multiple discriminant analysis. The degree to which members and different groups can be differentiated in terms of an array of discriminator variables is the essence of this technique. It may be very difficult in some instances to find a discriminating index number if the two samples have almost identical properties.

Discriminant analysis techniques are used to classify individuals into one of two or more alternative groups (or populations) on the basis of a set of measurements. The populations are known to be distinct, and each individual belongs to one of them. These techniques can also be used to identify which variables contribute to making the classification. Thus, as in regression analysis, we have two uses, prediction and description. The principal difference between a linear discriminant function and an ordinary linear regression function arises from the nature of the dependent variable. A linear regression function uses values of the dependent variable to determine a linear function that will estimate the values of the dependent variable, whereas the discriminant function possesses no such values or variable but uses instead a two-way classification of the data to determine the linear function. Consider for example, a set of k variables, $x_1, x_2, \ldots, x_k$, by means of which it is desired to discriminate between two groups of individuals. Letting

$$z = \lambda_1 x_1 + \lambda_2 x_2 + \ldots + \lambda_k x_k \tag{10.1}$$

represent a linear combination of these variables, the task is then to determine the λs by means of some criterion that will enable z to serve as an index for differentiating between members of the two groups. For example, for the purpose of simplification of the geometry of interpretation, consider two variables with n_1 and n_2 individuals, respectively, in two groups. The equation:

$$z = \lambda_1 x_1 + \lambda_2 x_2 \tag{10.2}$$

then represents a plane in three dimensions passing through the origin and having direction numbers λ_1, λ_2, and -1. The geometry of this discriminating process is shown in Fig. 10.1.

As another example, consider an archeologist who wishes to determine which of two possible indigenous groups created a particular statue found in an exploratory dig. Measurements are taken of several characteristics of the statue. It must now be decided whether these measurements are more likely to have come

Fig. 10.1. Example of a discriminating plane

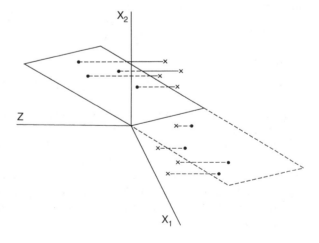

from the distribution characterizing the statues of one group or the other. The problem of classification is therefore to determine as best as is possible which group made the newly found statue on the basis of measurements obtained from statues whose identities are certain.

The measurements on the new statue may consist of a single observation, such as its height. However, we would then expect a low degree of accuracy in classifying the new statue since there may be quite a bit of overlap in the distribution of heights of statues from the two groups. If, on the other hand, the classification is based on several characteristics, we would have more confidence in the prediction. The discriminant analysis methods described in this chapter are multivariate techniques in the sense that they employ several measurements. This decision is made by determining whether the sample's characteristics are more similar to those of one population or another.

As another example, consider a geologist who wishes to decide to which formation a particular rock sample or water sample belongs. The decision is made by determining whether sample characteristics are more similar to those of formation 1, an arkose, or formation 2, a graywacke. Information on these two groups, available in past geological studies would include factors such as color, composition, texture, grain size, and other measurements.

10.4
Parameters for Classification

The parameters of interest in the discriminant model are: (1) the Mahalanobis distance that is adjusted for the variances and covariances of the responses, (2) classification probabilities i.e., the probability of sample assignment (3) Anderson's criterion, if calculated, and (4) classification or discriminant functions associated with groups. These parameters are discussed below.

10.5
Jackknifing Classification

Because bias enters classification when the coefficients used to assign a case to a group are derived, in part from the case, a jackknifing procedure is often used. In this procedure, the data from the case are left out when the coefficients used to assign it to a group are computed. Thus each case has a set of coefficients that are developed from all other cases, and the jackknifed classification gives a more reliable estimate of the ability of predictors to separate groups. When the procedure is used with all predictors forced into the equation, bias in classification is eliminated. However in the stepwise procedure where all predictors may not enter the equation, bias is reduced.

10.6
Similarities with Other Methods

Multiple discriminant analysis has a number of relationships with other multivariate techniques including regression analysis, factor analysis, and canonical correlation. As an example, when the discrimination involves only two groups, multiple discriminant analysis and multiple regression analysis are essentially the same thing, except that multiple discriminant analysis has a binary or dummy dependent variable in this case. In linear regression, the hypothesis focuses on the following question: what is the maximum combined linear combination of a cluster of variables with a dependent variable? In canonical correlation analysis, the question becomes: what is the maximum linear relationship, or relationships, between two clusters of variables. In discriminant analysis, the question becomes: what is the maximum linear relationship, or relationships, of a cluster of variables with a variable that is divided into categories.

In statistical methods ranging from analysis of variance through correlation, regression, and factor analysis, the aim is to account for variance reflected in relationships among variables. In multiple discriminant, factor analysis, and canonical correlation, the known relationships among individual variables are transformed to another configuration. In factor analysis, the larger array of relationships is reduced to a smaller subset called "factors". In canonical correlation, clusters are interrelated in terms of canonical variates. In multiple discriminant analysis, the relationship is characterized as discriminant functions. As in regression analysis, factor analysis, or canonical correlation, an attempt is made to interpret the clusters in terms of coefficients that indicate the degree of relationship of each of our original variables with the cluster. In the case of multiple discriminant analysis, each discriminant function is interpreted in terms of the various weightings of the original variables with that function.

The outcomes of various methods are however different in what is revealed. The factors in factor analysis take into account as much of the common variance in an overall matrix of interrelationships as is possible under the terms of the particular solution. In canonical correlation, the canonical variates give the maximum linear relationship of two clusters of variables. A discriminant function, like a canonical variate or a factor, may reflect different aspects of the same patterns of common variance among an array of variables but will not necessarily portray identical patterns.

10.7
Discriminant Function Coefficients

Both standardized and unstandardized discriminant function coefficients are used in the interpretation of discriminant functions as are discriminant scores. Standardized coefficients have the benefit of equalizing measurement scales and are the basis for interpreting relative contributions of discriminant variables to a function. They can be used to convert discriminant scores if the former are first standardized (which they are usually not). Unstandardized coefficients and the constant, like regression coefficients, are a basis for calculating discriminant scores from raw (unstandardized) scores of the discriminating variables. Because they reflect the differences of measurement scale, they are not an accurate basis for assessing relative contribution of discriminating variables to a discriminant function. In situations, particularly when multiple discriminant analysis is being used to classify new cases, based on known relationships between discriminating variables and groups, the equation involving unstandardized coefficients can be used in practical ways to show the overall relation as well as being a practical guide for making predictions.

10.8
Discriminant Scores and Probabilities

Discriminant scores are standardized scores with a mean of zero (standard score representation of overall mean of the combined groups on the discriminant score on that function) and a standard deviation of one. If the centroid of one group is −1.0, we are able to estimate that 34% of the cases in the overall distribution fall between the mean of that group and the grand mean of the overall group. If the centroid of the second group is +0.5, then we can deduce that approximately 53% of the cases fall between the two group means on that discriminant function.

When the correlation of a discriminant function is not perfect, i.e., the canonical correlation does not equal one, this implies that some error will be reflected in the assignment of cases to groups. This error may be considered as error in the body of the data, or error in assigning new cases. Error is defined when a case is assigned to the wrong group, where previously it was assigned to another group. Error is reflected in the overlap of distributions around different centroids and the higher the probability of overlap, the higher the likely error. It is a good idea to plot distributions and their overlap when computer analyses are available.

In summary, multiple discriminant analysis provides for the differentiation of single-variable groups or categories on the basis of relations with an array of discriminating variables. The calculation seeks to identify maximum multiple linear relations of discriminating variables with groups. Each set of relations is called a discriminant function. For each discriminant function, the calculations provide: (1) the canonical correlation of that function with the groups; (2) standardized discriminant function coefficients, which reflect the relative contributions of discriminating variables to that function; (3) unstandardized discriminant function coefficients and a constant, which may be used to convert raw scores on discriminating variables to discriminant function scores; and (4) centroids, or the average discriminant function scores for each of the groups. Multiple discriminant analysis is used both to describe the differentiation of groups, based on discriminating variables, as well as a basis for classifying new cases into likely groups. Discriminant function scores are standardized and thus are interpretable in terms of distributional probabilities. A good gauge of error is the comparison of prediction of group assignments with known cases (Williams 1979).

10.9
Discriminant Analysis Procedure

Discriminant analysis is concerned with separating distinct sets of objects mathematically, and with allocating new objects to previously defined groups using an allocator or discriminant function such as that of Fisher's method (1936). A common covariance matrix of groups is assumed but is often violated or not tested for. The linear discriminant function (Fisher 1936) is given by the equation:

$$Z = a_1 X_1 + a_2 X_2 + \ldots + a_p X_p, \tag{10.3}$$

where the means of groups are denoted by Z_1, Z_2, etc., and the pooled sample variance as S_z.

The linear discriminant function transforms an original set of measurements in a sample space into a single discriminant score; that score, or a transformed variable, represents the sample position along a line defined by the linear discriminant function (Davis 1973). The discriminant function is an important way to collapse a multivariate problem down to a problem which has only one variable. The discriminant function analysis is thus a means of finding a transform which gives the minimum ratio of the difference between a pair of group multivariate means to the multivariate variance within the two groups. Regression is one way to find linear discriminant functions.

To measure how far apart two groups are, we compute the Mahalanobis Distance according to:

$$D^* = \frac{(Z_1 - Z_2)^2.}{S_z} \tag{10.4}$$

Both Z_i in Eq. (10.4) and D^* are functions of the group means and the pooled variances and covariances of the variables. A larger value of D^* indicates that it is easier to discriminate between two groups. Posterior probabilities are used to express the likelihood of belonging to one group or another and is also an important measure of the ability to discriminate individuals based on size of probability of belonging to both groups.

10.10
Numerical Example on Ground-Water Sources

A discriminant analysis of ground-water sources was completed to investigate water quality. Steinhorst and Williams (1985) used several multivariate procedures to analyze ground-water sources. The procedures used in their paper included cluster analysis, MANOVA, canonical analysis, and discriminant analysis. From this study, the application of discriminant analysis will be looked at in greater detail. They used canonical variate analysis, not to be confused with canonical correlation analysis, to determine discriminants of the defined groups. They used discriminant analysis to achieve the objective of setting up a rule, based on measurements from individual samples. Canonical maps were made that showed the discrimination of the various water groups (Fig. 10.2). They determined posterior probabilities, reclassified the cluster group data, and classified unknown samples into defined groups as shown in Tables 10.1 and 10.2. The null hypothesis (Ho) in the MANOVA is that there is, in fact, no significant difference between any two groups of samples, based on true mean vector in groups.

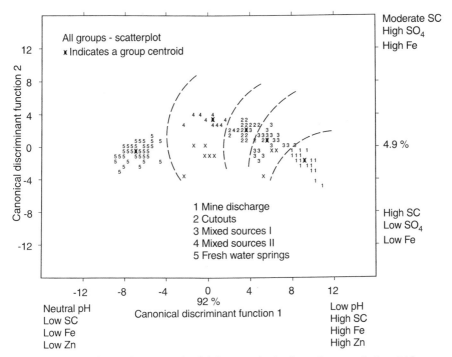

Fig. 10.2. Canonical map for case 1 (molybdenum mine), where the canonical variables are weighted sums of the seven original variables. (Steinhorst and Williams 1985) (Reprinted with kind permission of Water Resources Research. Copyright 1985)

Table 10.1. Summmary of MANOVA and canonical analysis for case 1 – molybedenum mine quality water. (Modified from Steinhorst and Williams 1985) (Reprinted with permission from Water Resources Research. Copyright 1985)

Step	Action Entered/Removed	Variables in	Wilks' λ	Significance
1	Log Zn	1	0.061	0.000
2	SC	2	0.021	0.000
3	Log Fe	3	0.008	0.000
4	pH	4	0.004	0.000
5	Log SO4	5	0.003	0.000
6	Log Mn	6	0.002	0.000
7	DS	7	0.002	0.000

Table 10.1 (continued)

Group	F-Statistic* and significance between pairs of groups			
	1	2	3	4
2	81.93*			
	0.00			
3	42.49*	33.21*		
	0.00	0.00		
4	124.80*	26.13*	46.57*	
	0.00	0.00	0.00	
5	635.06*	257.78*	439.29*	98.99*
	0.00	0.00	0.00	0.00

Canonical discriminant functions

Function	Eigenvalue	Variance percent	Cumulative percent	Canonical correlation
1	45.86	92.06	92.06	0.989
2	2.44	4.89	96.96	0.842
3	1.29	2.58	99.54	0.750
4	0.23	0.46	100.00	0.433

Standardized canonical discriminant function coefficients
Function

Variable	1	2	3	4
pH	−0.462	0.450	0.524	0.274
SC	0.521	−0.650	0.552	0.106
DS	−0.218	−0.409	0.349	−0.273
Log SO4	0.223	0.811	−0.200	−0.317
Log Fe	0.427	0.622	0.274	−0.362
Log Mn	0.149	0.237	−0.097	0.929
Log Zn	0.533	−0.200	−0.410	−0.084

Canonical discriminant functions evaluated at group means
Group   Function

Group	1	2	3	4
1	9.36	−2.22	0.42	0.47
2	3.44	1.93	2.20	−0.30
3	5.73	0.44	−1.49	−0.62
4	0.20	3.16	−1.20	1.11
5	−7.23	−0.61	0.02	−0.03

Each F-statistic has 7 and 132 degrees of freedom. SC, Specific electrical conductance at 25 °C; DS, total dissolved solids.

Table 10.2. Summary table for MANOVA and canonical analysis for case 2 – basalt flows and interbeds. (Modified from Steinhorst and Williams 1985) (Reprinted with permission from Water Resources Research. Copyright 1985)

Sep	Action Entered/Removed	Variables in	Wilks' λ	Significance
1	F	1	0.058	0.00
2	Ca	2	0.017	0.00
3	Cond	3	0.008	0.00
4	Cl	4	0.005	0.00
5	Mg	5	0.004	0.00
6	Na	6	0.003	0.00
7	pH	7	0.002	0.00

Goup	F-Statistics* and significances between pairs of groups		
	1	2	3
2	32.01*		
	0.00		
3	49.11*	127.02*	
	0.00	0.00	
4	136.64*	245.60*	33.90*
	0.00	0.00	0.00

Canonical discriminant functions

Function	Eigenvalue	Variance percent	Cumulative percent	Canonical correlation
1	44.77	90.56	90.56	0.989
2	3.42	6.92	97.48	0.880
3	1.25	2.52	100.00	0.744

Standardized canonical discriminant function coefficients

	Function 1	Function 2	Function 3
pH	0.211	−0.205	0.503
Cond.	1.056	0.185	−1.047
Cl	0.403	−0.333	0.742
F	0.418	0.371	0.268
Na	−1.030	0.509	0.167
Ca	0.011	0.776	−0.040
Mg	−0.927	0.466	0.785

Canonical discriminant functions evaluated at group means

Group	Function 1	Function 2	Function 3
1	4.028	−1.779	1.459
2	8.677	1.830	−0.647
3	−3.152	−1.976	−1.427
4	−7.763	1.564	0.499

Each F-statistic has 7 and 45 degrees of freedom. Cond; Specific electrical conductance at 25 °C.

10.11
Numerical Example on Hydrogeochemistry of Carbonate Terrains

A discriminant analysis was completed to investigate the geochemistry of carbonate waters and environments. Drake and Harmon (1973) applied discriminant analysis in studying the hydrochemical environments of carbonate terrains. They studied 162 chemical analyses of carbonate waters and tested the groupings using stepwise linear discriminant function analysis. They found that two parameters: (1) degree of calcite saturation and (2) equilibrium carbon dioxide partial pressure are sufficient to distinguish groups of waters. Six measured variables (temperature, calcium concentration, magnesium, bicarbonate, pH, and specific conductance) were used to determine equilibrium carbon dioxide partial pressure, the calcite saturation index, and the dolomite saturation index. Initially, the linear discriminant function analysis was done on the calculated variables and then for comparison analysis was done on the original variables. At each step, the variable entered into the discrimination was the one that gave the largest improvement in the F-ratio of between-group variation to within-group variation. Figure 10.3 shows the geochemical relationships established for the six water types. The results of the discriminant analyses are shown in Table 10.3.

10.12
Numerical Example on Stream Water Chemistry

Discriminant analysis was applied to study the geochemistry of stream waters in glaciated terrains in southwestern New York. Phillips (1988) used discriminant

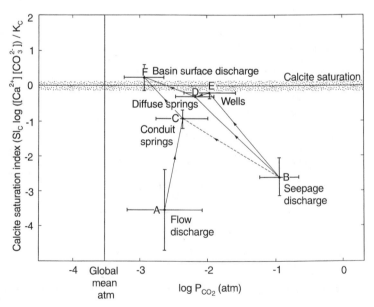

Fig. 10.3. Geochemical relationships for six water types. (Drake and Harmon 1973) (Reprinted with kind permission of Water Resources Research. Copyright 1973)

Table 10.3. (A) Variables entered into discriminant analysis, (B) Step at which groups are separate at the 0.005 level, and (C) classification matrices. (Modified from Drake and Harmon 1973) (Reprinted with kind permission of Water Resources Research. Copyright 1973)

(A)	Measured variables	Derives variables
Step 1	pH	SI_c
Step 2	pH, HCO_3	SI_c, $-\log P_{co_2}$
Step 3	pH, HCO_3, temp.	
Step 4	pH, HCO_3, temp., Ca^{2+}	
Step 5	pH, HCO_3, temp., Ca^{2+},	Mg^{2+}
Not entered at 0.005 level	Specific conductance	SI_d

(B) Measured variables	Groups A	B	C	D	E	F
A	0...					
B	1	0...				
C	1	1	0...			
D	1	1	2	0...		
E	1	1	2	2	0...	
F	1	1	1	1	1	0...

(C) Groups	A	B	C	D	E	F	No. Misclassified
Measured variables after two steps							
A	7	2	3	0	0	0	5
B	0	10	1	0	0	0	1
C	0	0	24	0	0	0	0
D	0	0	2	32	6	6	14
E	0	0	0	13	35	4	17
F	0	0	2	2	0	17	4
Constructed variables after two steps							
A	10	0	2	0	0	0	2
B	0	11	0	0	0	0	0
C	0	0	21	3	0	0	3
D	0	0	0	29	12	5	17
E	0	0	0	10	39	3	13
F	0	0	1	3	0	17	4

For the measured variables, 41 of 166 were misclassified (25%) and for the constructed variables, 39 of 166 were misclassified (23%)

analysis to study the relationship between glacial geology and stream water chemistry in an area receiving acid deposition. For seventy streams in southwestern New York that were sampled in June and again in December, stream water chemistry proved to be closely linked to glacial geology. He found that glacial geology of a watershed is well predicted by discriminant analysis of stream water chemistry data. Discriminant analysis into unglaciated and glaciated categories correctly classified 91–95% of all streams for the three sampling sites without jackknifing, and 90–93% with jackknifing (Table 10.4). The discriminant analysis also showed a similiarity between Kent drift sheet stream waters and Olean drift sheet stream waters because a high number (7 out of 10) of Olean drift sheet streams were classified as Kent drift sheet (Table 10.5).

10.13
Numerical Example on Oil Field Chemistry

Discriminant analysis was applied in a geological analysis of fluid characteristics in oil reservoirs in California. The results and analyses are briefly summarized. Gerrild and Lantz (1969) collected crude oil samples from sandstone in the Elk Hills, California petroleum reserve. A discriminant analysis was used to assign samples based on their chemistry. For 5 variables measured on 56 cases whose population assignment was certain, Fisher linear discriminants showed that the separation of three group means is fully explained in the two-dimensional discriminant space. The results of the study are shown on Fig. 10.4.

Table 10.4. Discriminant analysis of unglaciated (U) and combined glaciated (G) area samples. (Modified from Phillips 1988) (Reprinted with kind permission from Elsevier Science – NL, Sara Burgerhartstraat 25, 1055 KV Amsterdam, the Netherlands. Copyright 1988)

Date	Area	N	Classification					
			Unjackknifed			Jackknifed		
			Percent correct	No. class. into area		Percent correct	No. Class. into area	
				U	G		U	G
June 25	U	25	100	25	0	100	25	0
	G	8	89	2	16	83	3	15
	Total	43	95	27	16	93	28	15
June 28	U	9	78	7	2	78	7	2
	G	18	100	0	18	100	0	18
	Total	27	93	7	20	93	7	20
Dec 7–8	U	33	94	31	2	94	31	2
	G	36	89	4	32	86	5	31
	Total	69	91	35	34	90	36	33

Table 10.5. Discriminant analysis of unglaciated area (U), Olean drift sheet (O), and Kent drift sheet (K). (Modified from Phillips 1988) (Reprinted with kind permission from Elsevier Science – NL, Sara Burgerhartstraat 25, 1055 KV Amsterdam, the Netherlands. Copyright 1988)

Date	Area	N	Classification							
			unjackknifed				Jackknifed			
			Percent correct	No. class. into area			Percent correct	No. class. into area		
				U	O	K		U	O	K
June 28	U	9	78	7	0	2	78	7	0	2
	O	6	50	0	3	3	33	0	2	4
	K	12	83	0	2	10	83	0	2	10
	Total	27	74	7	5	15	70	7	4	16
Dec 7-8	U	33	94	31	2	0	91	30	3	0
	O	24	79	2	19	3	75	3	18	3
	K	12	83	0	2	10	67	1	3	8
	Total	69	87	33	23	13	81	34	24	11

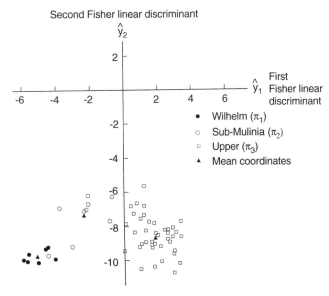

Fig. 10.4. Crude oil samples mapped in discriminant space. (Gerrild and Lantz, 1969)

10.14
Summary

Discriminant analysis is very important when investigating the characteristics of different sample populations and assigning a classification to individual samples.

10.15
Supplemental Reading

Afifi AA, Azen SP (1972) Statistical analysis – a computer oriented approach. Academic Press, New York
Davis JC (1973) Statistics and data analysis in geology. John Wiley, New York
Tabachnick BG, Fidell LS (1989) Using multivariate statistics. Harper and Row, New York

Individual-Directed Techniques not Based on Normal Distribution Assumptions

Cluster Analysis

11.1
Concept

The objective of cluster analysis is to separate the groups based on measured characteristics so as to maximize distance between groups.

11.2
Definitions

Distance Coefficient. A mathematically defined measure of the separation between two objects.

Dendrogram. A tree diagram showing similarity or connectivity between objects and clusters.

Correlation Coefficient. A measure of similarity wherein highest correlations or similarities are linked first in clusters.

11.3
Overview

Cluster analysis is a separate and useful technique for grouping individuals or objects into unknown groups. It differs from other methods of classification, such as discriminant analysis in that in cluster analysis the number and characteristics of the groups are to be derived from the data and are not usually known prior to the analysis, thus it is not a true classification technique.

Cluster analysis has been used for decades in the area of taxonomy in the biological sciences. In taxonomy, living things are classified into arbitrary groups on the basis of their characteristics. The classification proceeds from the most general to the most specific, in steps.

Cluster analysis has been used in medicine to assign patients to specific diagnostic categories on the basis of their presenting symptoms and signs. In particular, cluster analysis has been used in classification in medicine, archaeology, geology, hydrology, and anthropology. Cluster analysis may be R-mode or R-type (classification of variables) and Q-mode or Q-type (classification of samples or objects).

The parameters of most interest in cluster analysis are (1) the distance measure – the Mahalanobis distance – that is a value adjusted for the variances and covariances of the responses, or (2) other measures of similarity or dissimilarity in matrix form.

11.4
Cluster Analysis Procedure

The matrix of data observations (nxp) is transposed to an nxn matrix to determine measures of similarity among the observations, i.e., giving rise to a matrix of distances or similarities between samples. Cluster analysis graphically defines a similarity matrix with a dendrogram in most simple cases. A clustering algorithm is used to find mutually exclusive groups or hierarchical clusters, and a comparison of the clusters is made that attempts to maximize the distance between clusters. One measure of dissimilarity is the previously given Mahalanobis D* or other measures such as the "centroid distance" or group average. Most techniques for analysis are straightforward and give similar results. Of the hierarchical methods, single link methods have proven very useful. The underlying principles for selecting criteria tests for best classification depends on the fundamental relationship that total variation (T) is composed of the variation between groups (B) plus the variation within groups (W) :

$$T = B + W. \tag{11.1}$$

One criterion is the maximizing of the logarithm of the ratio between the determinant of:

$$T \text{ and } B \times \log |T|/|W|, \tag{11.2}$$

$$\text{where } (T/B) = 1 + W, \tag{11.3}$$

so that maximizing $\log (T)/(W)$ is the requirement. A second criterion uses Mahalanobis D^2, a distance function in which distance between groups is compared with distance within groups. Additionally, when groups are requested in clustering, the criteria may be plotted to formulate a clear set of groupings. A procedural outline is shown in Fig. 11.1.

11.5
Numerical Example on Properties of Mineralized Waters

Cluster analysis was performed on data collected to study mineralized waters in Montana, and a short summary of results is provided. Williams (1982) analyzed the hydraulic connections between the surface of a mountain and internal mineralized sources for Mount Emmons near Crested Butte, Mountana. The objective of this study was to delineate preferential flow pathways based on data from seeps and springs in the area. One hundred fifty-four analyses of water samples collected on Mount Emmons between 1 August, 1979, and 15 January, 1980 were included for study. The parameters studied were pH, specific conductance (SC), in μmhos/cm; dissolved solids (DS), in mg/l, sulfate (SO^4), in mg/l, iron (Fe), in mg/l, manganese (Mn), in mg/l, and zinc (Zn), in mg/l. A number of multivariate procedures were used in this study, among them cluster analysis, MANOVA, and canonical and discriminant analysis. Using stepwise MANOVA, the single most discriminating variable was found to be zinc. By using a hierarchical clustering technique, cluster analysis in conjunction with canonical analyses of water quality data delineated those discharge points that contained flow emanating from a mineralized pyrite-rich zone.

The resulting clusters, modified by the results of the canonical analysis, along with fault-vein maps, identified springs which would be most likely affected by mining in the core of Mount Emmons (Williams 1982). Data for the MANOVA are shown in Tables 11.1 and 11.2. Figure 11.1 is a cluster diagram delineating sources. Canonical maps are plots of the canonical discriminant functions (Figs. 11.2 and 11.3).

Table 11.1. Summary of seven variable, five group canonical analysis. (Modified from Williams 1982) (Reprinted by permission of Ground Water. Copyright 1982. All rights reserved)

Step	Action Entered/Removed	Vars in	Wilks' λ	Significance
1	Log Zn	1	0.061	0.000
2	SC	2	0.021	0.000
3	Log Fe	3	0.008	0.000
4	pH	4	0.004	0.000
5	Log SO$_4$	5	0.003	0.000
6	Log Mn	6	0.002	0.000
7	DS	7	0.002	0.000

F-Statistics and significance between pairs of groups

Group	1	2	3	4
2	81.93			
	0.00			
3	42.49	33.21		
	0.00	0.00		
4	124.80	26.13	46.57	
	0.00	0.00	0.00	
5	635.06	257.78	439.29	98.99
	0.00	0.00	0.00	0.00

Canonical discriminant functions

Function	Eigenvalue	Variance Percent	Cumulative Percent	Canonical Correlation
1	45.86	92.06	92.06	0.989
2	2.44	4.89	96.96	0.842
3	1.29	2.58	99.54	0.750
4	0.23	0.46	100.00	0.433

134

Table 11.1 (continued)

Standardized canonical discriminant function coefficients functions				
Variable	1	2	3	4
pH	−0.462	0.450	0.524	0.274
SC	0.521	−0.650	0.552	0.106
DS	−0.218	−0.409	0.349	−0.273
Log SO$_4$	0.223	0.811	−0.200	−0.317
Log Fe	0.427	0.622	0.274	−0.362
Log Mn	0.149	0.237	−0.097	0.929
Log Zn	0.533	−0.200	−0.410	−0.084

Canonical discriminant functions evaluated at group means				
Group	Function			
	1	2	3	4
1	9.36	−2.22	0.42	0.47
2	3.44	1.93	2.20	−0.30
3	5.73	0.44	−1.49	−0.62
4	0.20	3.16	−1.20	1.11
5	−7.23	−0.61	0.02	−0.03

Each F-statistic has 7 and 132 degrees of freedom. SC, specific electrical conductance at 25 °C; DS, total dissolved solids.

Table 11.2. Summary of seven variable canonical analysis for two clusters (groups). (After Williams 1982. Reprinted by permission of Ground Water. Copyright 1982)

Step	Action Entered/Removed	Variables in	Wilks' λ	Significance
1	Log SO$_4$	1	0.1975	0.00
2	Log Fe	2	0.1710	0.00
3	SC	3	0.1487	0.00
4	pH	4	0.1393	0.00
5	Log Mn	5	0.1319	0.00
6	Log Zn	6	0.1292	0.00

Function	Eigenvalue	Percent of variance	Cumulative %	Canonical correlation
1	6.739	100	100	0.9332

Function	After function	Wilks' λ	Chi Squared	D.F.	Significance
1	0	0.1292	282.38	6	0.0

Standardized canonical discriminant function coefficients			
Function 1		Function 1	
pH	−0.02504	Log Fe	0.65745
SC	0.65186	Log Mn	−0.69671
Log SO$_4$	0.22609	Log Zn	0.32647

Canonical discriminant functions evaluated at group means (group centroid)	
Goup	Function 1
1	2.632
2	−2.524

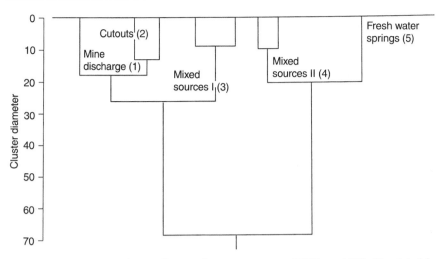

Fig. 11.1. Seven variable cluster diagram for water sources. (Williams 1982) (Reprinted by permission of Ground Water. Copyright 1982. All rights reserved)

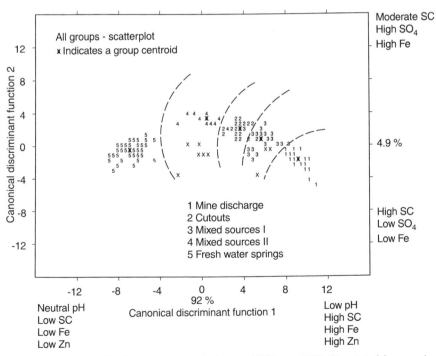

Fig. 11.2. Seven variable, five group canonical map. (Williams 1982) (Reprinted by permission of Ground Water. Copyright 1982. All rights reserved)

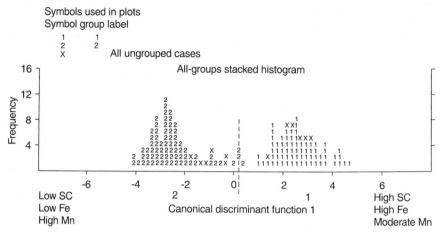

Fig. 11.3. Seven variable, two group canonical map. (Williams 1982) (Reprinted by permission of Ground Water. Copyright 1982. All rights reserved)

11.6
Numerical Example on Stream Water Quality

Variable (R-mode) and sample (Q-mode) cluster analyses were performed on data collected for hydrological study of stream water quality in western New York. Bloomfield (1976) applied cluster analysis to ascertain differences in stream water quality. Cluster analysis was used to examine the spatial patterns, and stream water quality of 44 watersheds in the Genesee River basin in western New York State. Nine groups of watersheds and seven groups of water quality variables were identified, and three clusters of state variables were produced. Cluster analysis of samples yielded subsets representing runoff events, recessional periods, and base flows (Bloomfield 1976). The examination of the geographic distribution of the data is done in the Q-mode analysis and the examination of variable groups is done in the R-mode analysis. The Q-mode dendrogram for Genesee data is shown in Fig. 11.4. The R-mode dendrogram of Genesee data is shown in Fig. 11.5. The Q-mode dendrogram for Mill Creek data is shown in Fig. 11.6 and the R-mode dendrogram is in Fig. 11.7. The cluster patterns can be easily observed in the dendrograms.

11.7
Variable Clustering

A helpful way of analyzing multivariate data is to use a profile diagram which uses standardized or unstandardized data plotted on a graph. The value of variables measured on the same object are connected by a straight line between segments. An example of a profile diagram is shown in Fig. 11.8. The x-axis is the

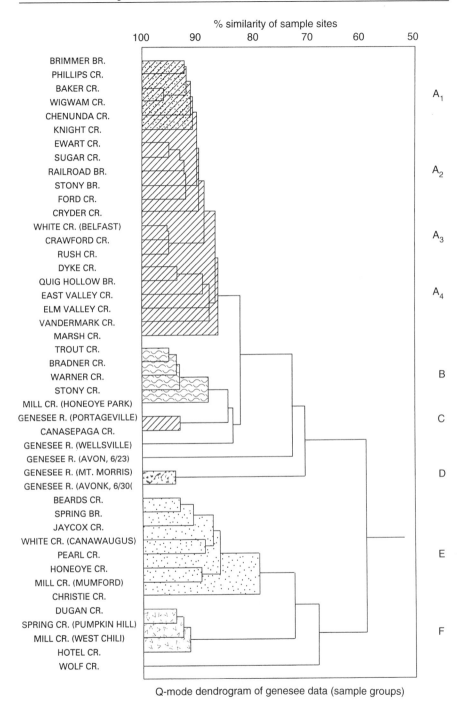

Q-mode dendrogram of genesee data (sample groups)

Fig. 11.4. Q-mode dendrogram for Genesee data. (Bloomfield 1976)

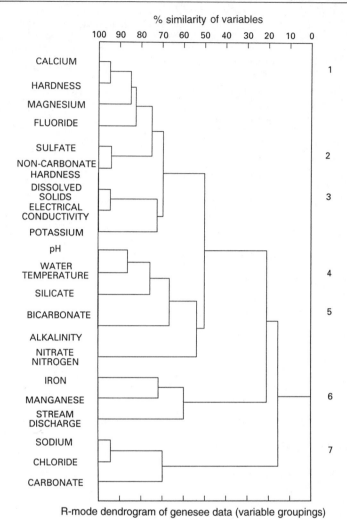

Fig. 11.5. R-mode dendrogram for Genesee data. (Bloomfield 1976)

variable label, i. e., varable 1, variable 2, ... , variable n. The vertical spacing for a selected variable is used to show similarities between individual samples or objects.

Variable cluster analysis is another useful technique for looking at variable relationships. It makes use of the correlation coefficient. A cluster diagram is derived, an example of which is shown in Fig. 11.9. Parks (1966) analyzed data on 200 sediment samples to derive this diagram.

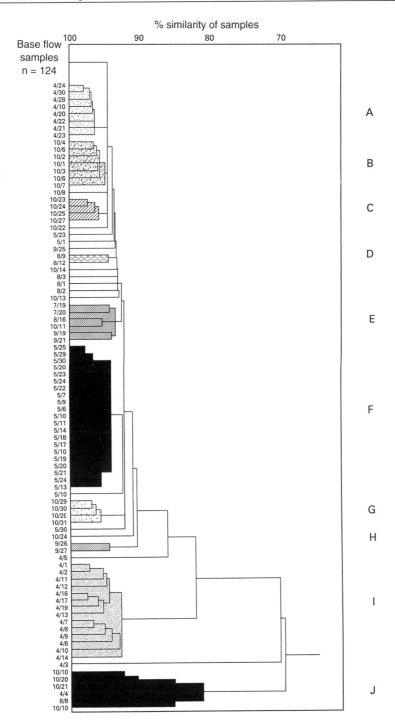

Fig. 11.6. Q-mode dendrogram for Mill Creek data. (Bloomfield 1976)

% similarity of variables

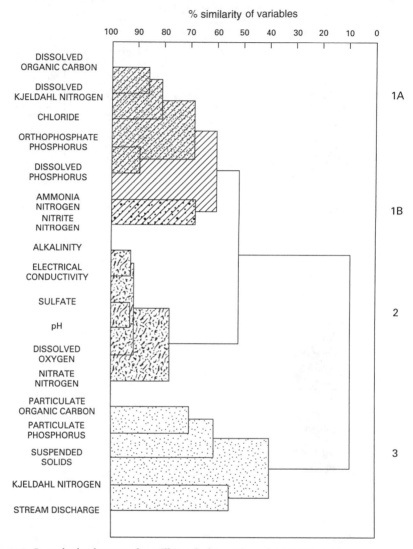

Fig. 11.7. R-mode dendrogram for Mill Creek data. (Bloomfield 1976)

11.8
Integrated Cluster Analysis

A study was completed that used other statistical techniques with the clustering algorithm to ascertain differences in rock properties and as such is called integrated cluster analysis. A short summary of results is provided.

The procedure for cluster analysis has been previously discussed and the reader is referred to Section 11.3 for an introduction and general discussion

Fig.11.8. Profile diagram of measurements on three individual rocks showing similarities in variables. (After Davis 1986). (Reprinted by permission of John Wiley and Sons, Inc. Copyright 1986. All rights reserved)

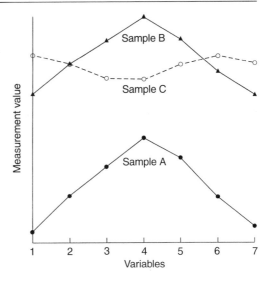

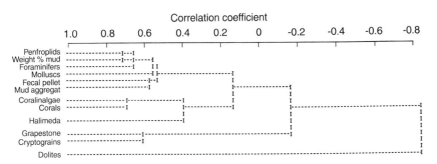

Fig. 11.9. Cluster diagram for variables using the correlation coefficient. (After Parks 1966) (Reprinted from Geology with kind permission of the University of Chicago Press. Copyright 1966. All rights reserved)

of concepts. The concept of integrated analysis is used to show that cluster analysis has been used synergistically with other statistical methods such as the Kruskal-Wallis one-way analysis of variance (AOV). The Kruskal-Wallis AOV is a nonparametric technique for investigating variable groups.

11.9
Numerical Example on Carbonate Rock Data

Cluster analysis and the Kruskal-Wallis test were applied in the study of rock properties in central Pennsylvania. The Bellefonte dolomite, Axemann limestone, Nittany dolomite, and Stonehenge limestone were selected for study of geohydrological factors influencing porosity and permeability characteristics of carbonate rocks in central Pennsylvania (Brown 1977, 1993).

A suite of 32 samples were collected and studied to determine the potential of carbonate rock aquifers of the Lower Ordovician Beekmantown group in central Pennsylvania. The aquifer potential was described by looking at sample groups using factor analysis, cluster analysis, and a suite of univariate tests. Cluster analysis was used on samples from the Nittany (N), Stonehenge Limestone (S), Axemann limestone (A), and Bellefonte dolomite (B) and the number of meaningful groups (based on different properties demonstrated) were determined (Fig. 11.10). The original data are given in Table 11.3. Table 11.4 shows the result of clustering of the samples. The significant variables that were found represent-

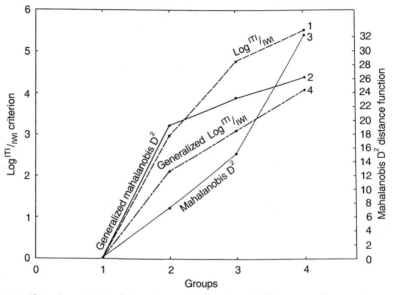

Fig. 11.10. Clustering criteria plots to determine number of different aquifer groupings

Table 11.3. Rock properties data from laboratory analysis of 31 samples of Beekmantown carbonates, central Pennsylvania. (Brown 1977)

Sample No.	Bulk density (g/cc)	Porosity (%)	Log permeability	Permeability (millidarcies)
S-1	2.69	0.83	−1.2218	0.06
S-5	2.64	0.59	−1.6989	0.02
S-8	2.64	0.67	−1.6989	0.02
S-3	2.63	0.50	−1.6989	0.02
S-7	2.62	1.10	−1.2218	0.06
N-6	2.72	2.10	−1.6989	0.02
N-7	2.74	2.80	−1.2218	0.06
N-8	2.66	4.60	−0.7213	0.19
N-9	2.69	3.50	−1.5228	0.03
N-10	1.70	8.60	−1.3010	0.05

Table 11.3 (continued)

Sample No.	Bulk density (g/cc)	Porosity (%)	Log permeability	Permeability (millidarcies)
N-11	2.73	0.81	−1.6989	0.02
N-12	2.74	1.40	−2.0000	0.01
N-13	2.68	2.70	−1.6989	0.02
A-14	2.75	1.20	−1.5229	0.03
A-15	2.71	0.91	−1.6989	0.02
A-16	2.62	1.40	−1.3979	0.04
A-17	2.72	3.00	−1.3010	0.05
A-18	2.61	1.30	−1.5228	0.03
A-19	2.63	0.18	−2.0458	0.009
A-20	2.62	5.40	−1.6989	0.02
A-21	2.65	0.30	−2.3010	0.05
A-22	2.62	0.56	−1.6989	0.02
B-23	2.76	0.93	−2.0000	0.01
B-24	2.78	0.23	−1.6989	0.02
B-25	2.76	1.50	−1.6989	0.02
B-26	2.77	0.74	−1.6989	0.02
B-27	2.74	0.68	−2.0000	0.01
B-28	2.76	1.70	−1.6989	0.02
B-29	2.71	2.70	−1.5228	0.03
B-30	2.66	2.90	−1.6989	0.02
B-31	2.84	1.8	−1.6989	0.02

S, Stonehenge limestone; N, Nittany dolomite; A, Axemann limestone; B, Bellfonte dolomite.

Table 11.4. Parameters derived using Log $|T|\backslash|W|$ criterion for cluster analysis (Brown 1977)

Requested No. of Groups	Log $|T|\backslash|W|$	Generalized distance	Characteristics of requested groups
2	2.997	19.03	Group 1 (dolomites and dolomitic limestones with largest porosity) Group 2 (limestones, low magnesium content and low porosity)
3*	4.612	23.14	Group 1 (high magnesium limestones, dolomites, with highest insoluble residue values and medium to low porosity) Group 2 (same as above) Group 3 (high magnesium limestones, dolomites, low to medium insoluble residues, medium to highest porosity)
4	5.561	26.36	*Group 2 (dissected) with/and 3 from 3* above

ed by the samples are given in Table 11.4. Aquifer potential properties were uniquely defined to the various sample groups as shown through the Kruskal-Wallis tests (Table 11.5).

The coefficient of variation was also examined as a measure of diversity (Table 11.6). Further examination of the samples from formations under study indicate that the aquifers under study are indeed different in terms of aquifer productivity (Tables 11.7 and 11.8). Cluster analysis showed that samples could be divided into three significantly different groups which contain samples of

Table 11.5. Results of Kruskal-Wallis test for porosity

Formation	No. obs.	Median	Ave. Rank	Z-Value
Stonehenge	5	0.67	8.4	−2.04
Nittany	8	2.75	23.2	2.62
Axemann	9	1.20	14.1	−0.76
Bellefonte	9	1.50	15.7	−0.11

H-statistic = 9.00, d.f. = 3, prob. = 0.030
H-statistic = 9.00, d.f. = 3, prob. = 0.030 (adjusted for ties)
Critical H = 7.815

Result: Reject Ho: of equal means.
　　　Do not reject Ha: at least one is different in terms of porosity means.

Table 11.6. Coefficient of variation for porosity

Formation	Mean	Standard deviation	Coefficient of variation
Stonehenge ls.	0.74	0.236	0.32
Nittany dol.	3.31	2.44	0.74
Axemann ls.	1.58	1.66	1.05
Bellefonte dol.	1.72	0.937	0.54

Table 11.7. Comparison of well productivity values (in 10^3 gpm/ft/ft) for wells in central Pennsylvania. (Siddiqui 1969)

Rock type	N	Minimum	Maximum	Median	Geometric mean
Shale	10	0.56	85.65	3.10	3.98
Bellefonte dol.	22	0.11	180.14	4.45	3.58
Limestone	11	0.31	323.73	7.89	11.09
Nittany dol.	15	0.85	4360.68	68.30	79.43
Upper Sandy dol.	22	0.70	2202.56	122.35	123.30

gmp, gallons per minute.

Table 11.8. Productivity of wells (in gpm/ft/ft) for central Pennsylvania aquifers. (Siddiqui 1969)

Nittany dolomite (15 wells)	Bellefonte dolomite (22 wells)	
52.64	1.36	
6.68	1.90	
2271.71	0.84	
224.09	0.11	
68.30	12.43	
3.01	0.28	
491.45	0.14	
8.43	15.66	
0.85	9.04	
3621.06	1.26	
3269.73	7.00	
4360.68	0.37	
	47.76	
109.37	14.86	
41.76	14.10	
	8.01	
Total 14 545	0.41	
Median = 68.30	1.43	
	56.89	
	180.14	Total 429.88
	0.13	Median = 4.45

gpm, gallons per minute

distinctly different porosity and composition. The Kruskal-Wallis test and coefficient of variation were also used to look at other significant variables that were separated according to formation. The dolomite formations were found to have higher values of porosity and permeability than the limestone formations. The productivity of wells in selected formations was previously analysed by Siddiqui (1969) and comparisons made. He found that the upper sandy dolomite of the Gatesburg formation and the Nittany dolomite had the largest median and geometric mean values of the rock types studied (Table 11.7), leading one to conclude that the Gatesburg Formation and Nittany dolomite represent the best aquifers based on well productivity of those studied.

11.10
Summary

The cluster analysis method has been applied both to the analysis of variables and objects (individuals) in this chapter. Many other examples have been completed and can be found in the literature.

11.11
Supplemental Reading

Afifi AA, Azen SP (1972) Statistical analysis – a computer oriented approach. Academic Press, New York

Afifi AA, Clark V (1990) Computer-aided multivariate analysis, 2nd edn. Van Nostrand Reinhold, New York

Bloomfield JA (1976) The application of cluster analysis to stream water quality data, in Proceedings of conference on environmental modeling and simulation. US Environmental Protection Agency, Washington DC, p. 683–690

Davis JC (1973) Statistics and data analysis in geology. John Wiley, New York

Davis JC (1986) Statistics and data analysis in geology, 2nd edn. John Wiley, New York

Tabachnick BG, Fidell LS (1989) Using multivariate statistics. Harper and Row, New York

Williams RE (1982) Statistical identification of hydraulic connections between the surface of a mountain and internal mineralized sources. Ground Water 20(4): 466–478

CHAPTER 12

Multiple Logistic Regression

12.1
Concept

The multiple logistic regression equation is based on the premise that the natural log of odds (logit) is linearly related to independent variables. The logit equation is the same as for the discriminant function and multiple regression equation with the dependent variable as the natural log of odds.

12.2
Overview

Ordered categorical variables arise frequently in a wide variety of experimental studies in the sciences. Two classes of models for the analysis of ordered categorical data are in more common use. These models are: (1) log – linear models and (2) proportional – odds models (called logit models). Like linear models, logit models distinguish between response and explanatory variables, while log – linear models do not. In addition, logit models do not require arbitrary scores to be assigned to the categories of the response variable. Log-linear models use the logarithm of cell frequencies in a linear model and are similar to chi-square goodness-of-fit tests on cell frequencies.

12.3
Logistic Regression

The study of binary response and explanatory variables is better done using a logistic regression which determines a curvilinear model relationship, rather than a linear probability model. The linear probability model has a major structural defect that is related to the probability model requirements that values fall between zero and one. In some situations, logistic regression modeling is also preferred for analysis over discriminant analysis when multivariate normality is not apparent, and an individual is to be classified into one of two populations. Thus it is an alternative to discriminant analysis, but if the data are multivariate normal, a smaller set of data is required by discriminant analysis to achieve the same precision as the logistic regression model.

These techniques have been widely used in the medical field and private industry to predict success or failure of a process based on several measured variables. A complete summary of logistic regression can be found in Agresti (1990).

12.4
Log-Linear Models

A special class of statistical techniques, called log – linear models, is used for the analysis of categorical data. These models are useful for uncovering relationships among variables in a multiway cross-tabulation. Log – linear models are very similiar to multiple regression, but in log – linear models all the variables for classification are independent variables and the dependent variable is the number of cases in a cell of the crosstabulation (Norusis 1985). The reader is referred to Norusis (1985) for examples of special application.

When studying relations between categorical variables where one is dependent while the others are independent variables, a log-linear analysis can be used and the analysis is a "logit" model. Logit models are specific members of the class of generalized linear models for ordinal data, which are specified by some link function relating the dependent variable to a linear combination of the independent variables. Logit models assume an underlying logistic distribution (logit link function), which allows the model parameters to be interpreted in terms of the odds ratios, and are multiple regression-like applications of multiway frequency analysis.

If the dependent variable has only two categories, either multiple regression/analysis of variance or logit analysis are appropriate analyses, wherein the results of logit analysis are usually close to those of multiple regression/analysis of variance with a dichotomous dependent variable.

Once a suitable model is identified and the parameters for each cell in a contingency table is computed, the parameters can be converted to odds (Tabachnick and Fidell 1989). In fact, logit refers to the interpretation of the parameters as the log of odds ratios. The general equation for converting a parameter to odds ratio is (Tabachnick and Fidell 1989):

$$\text{Odds ratio} = e^{2\lambda}, \qquad\qquad\qquad (12.1)$$

where λ = class (cell) parameter, and the log of odds is a linear relationship.

12.5
Probit Models

The alternative assumption of normality of data has also been considered in selected references listed and discussed previously. Probit (probit link function) models for ordinal data behave similarly to logit models because of the similarity of the normal and logistic distributions (Agresti 1990); however, since parameter interpretation is in terms of standard deviations instead of odds ratios, these models are more similar than logit models to linear models.

12.6
Log-Linear Models for Contingency Tables

Log – linear models for multidimensional contingency tables also describe association patterns among three or more categorical variables when several variables, all being discrete, are involved in the contingency tables. This procedure is also called a multiway frequency analysis in some instances. This topic is comprehensively covered in other texts and the reader is referred to Agresti (1990) and Tabachnick and Fidell (1989) for additional insight.

12.7
Numerical Example on Geochemical Data

A logistic regression analysis was applied to study the geochemical nature of organic solvent-trichloroethylene (TCE). Logistic regression also called logit regression has not been widely used in the geohydrologic sciences. One good discussion of logistic regression and its uses in water resources investigations is given in Helsel and Hirsch (1992). They define the method as transforming the estimated probabilities (p) into a continuous response variable with values possible from minus to plus infinity. The transformed response is predicted from one or more explanatory variables, and subsequently retransformed back to a value between zero and one (Helsel and Hirsch 1992). They define the model equations for multiple logistic regression with appropriate criteria tests. Tables 12.1 and 12.2 give the results of a logistic regression model analysis of TCE data (collected by Eckhardt et al. 1989), that included one- and two-variable models. A plot of the logistic regression line is shown in Fig. 12.1. To determine the significance of population density (POPDEN) as an explanatory variable, the

Table 12.1. Statistics for POPDEN model. (Helsel and Hirsch 1992) (Reprinted with kind permission of Elsevier Science-NL, Sara Burgerhartstraat 25, 1055 KV Amsterdam, The Netherlands. Copyright 1992. All rights reserved)

Explanatory variable	Estimate	Partial T-statistic	p-Value
Intercept	−2.80	−13.4	0.0001
Popden	0.226	8.33	0.000

Table 12.2. Statistics for the POPDEN plus SEWER model. (Helsel and Hirsch 1992) (Reprinted with kind permission of Elsevier Science-NL, Sara Burgerhartstraat 25, 1055 KV Amsterdam, The Netherlands. Copyright 1992. All rights reserved)

Explanatory variable	Estimate	Partial T-statistic	p-Value
Intercept	−3.24	−12.47	0.0001
Popden	0.13	4.07	0.0001
Sewer	1.54	4.94	0.0001

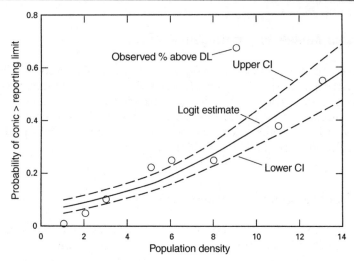

Fig. 12.1. Logistic regression line for TCE data, with percent detections observed for each population density. (Helsel and Hirsch 1992) (Reprinted with kind permission of Elsevier Scienc-NL, Sara Burgerhartstraat 25, 1055 KV Amsterdam. The Netherlands. Copyright 1992. All rights reserved)

likelihood ratio (lr) is computed by subtracting the log likelihood of this one variable model from that of the intercept-only model, and comparing to a chi-square distribution (wherein, $lr = 610.0 - 533.0 = 77.0$ with 1 df, resulting in a p-value of 0.0001 (Helsel and Hirsch 1992)). A second variable was added that related a sewered versus an unsewered area around the well under study. Table 12.2 shows that the logistic regression model for two variables was an improvement over the previous model with one variable.

The likelihood ratio test determined that the two-variable model is better than an intercept only model: wherein, $lr(0) = 610.0 - 506.3 = 103.7$ with 2 df, resulting in a p-value of 0.0001. Helsel and Hirsch (1992) also looked at the addition of other variables to the model, but results are not presented here.

12.8
Relation to Discriminant Analysis

Press and Wilson (1978) presented in their paper an explanation for choosing between logistic (logit) regression and discriminant analysis. In this paper, logistic regression is defined as the method of relating qualitative variables to other variables through a cdf (cumulative distribution function) functional form. Discriminant analysis is the method of classifying an observation into one of several populations. There are situations in which discriminant analysis and logistic regression are both appropriate, but the discriminant analysis is more powerful in the presence of multivariate normality of the data.

12.9
Multinomial Response Models

Some loglinear models do not distinguish between response and explanatory variables. Some logit models treat nominal response variables whereas others treat ordinal variables. Other models are designed for categorical response variables having more than two categories and are generalizations of logit models for binary responses. Some models use different transformations of response probabilities when ordinal data are involved. Like binary logit models for categorical variables, these models are equivalent to log – linear models for multi-way contingency tables. Multinomial response models explicitly treat one variable as a response and use a sampling model that fixes counts at combinations of levels of the explanatory variables. It assumes that the response counts at each combination have a multinomial distribution, and that multinomial counts at different combinations are independent. Examples in the biological field include studies on food chains of organisms. Other investigations are given in Agresti (1990), and these analyses and methods may be easily adapted to geohydrologic studies. Agresti (1990) also discusses the variations of logit and loglinear models for multinominal response data.

12.10
Summary

Several studies have been reviewed that apply logistic regression models to the data. These methods have been applied more widely in other scientific disciplines, but methods can be easily applied to geohydrologic data.

12.11
Supplemental Reading

Agresti A (1990) Categorical data analysis. John Wiley, New York

Other Approaches to Explore Multivariate Data

CHAPTER 13

Coefficient of Variation

13.1
Concept and Procedure

If the absolute dispersion is defined as the standard deviation, and the average is the mean, the relative dispersion is called the coefficient of variation (CV) or coefficient of dispersion. The relationship between mean and dispersion is very important in the geosciences and is expressed by the coefficient of variation according to:

$$CV\% = 100\,\sigma/mean, \tag{13.1}$$

where σ = standard deviation. The coefficient of variation is attractive as a statistical tool because it apparently permits the comparison of variates free from scale effects; i.e., it is dimensionless. However, it has appropriate meaning only if the data achieve ratio scale. The coefficient of variation can be plotted as a graph to compare data. A CV exceeding say about 30 percent is often indicative of problems in the data or that the experiment is out of control. Variates with a mean less than unity also provide spurious results and the coefficient of variation will be very large and often meaningless.

13.2
Numerical Example on Geochemical Data of Karst Aquifer

The coefficient of variation was applied to study the geochemistry of waters in a karst aquifer in Kentucky. Scanlon (1990) studied the relationship between ground-water contamination and major ion chemistry in a karst aquifer. Ground-water contamination was examined within the Inner Bluegrass karst region of central Kentucky where potential contaminant sources were varied. To evaluate controls on ground-water contamination, data on nitrate concentrations and indicator bacteria in water from wells and springs were compared with physical and chemical attributes of the ground-water system (Scanlon 1990). Tables 13.1 and 13.2 summarize the characteristics of water types and subtypes and coefficient of variation for each. Scanlon (1990) found that the occurrence and concentrations of nitrate and bacteria in ground-water correlate with the distribution of chemical water types in the aquifer.

Table 13.1. Summary of chemical characteristics of water types and subtypes. (Scanlon 1990) (Reprinted with kind permission of Elsevier Science-NL, Sara Burgerhartstraat 25, 1055 KV Amsterdam, The Netherlands. Copyright 1990. All rights reserved)

	Ca-Mg-HCO$_3$		Na-HCO$_3$	Na-Cl
Water subtypes:	Ca	Ca–Mg	–	–
No. samples:	108	77	10	4
No. wells/springs:	18	11	2	3
Ca				
Range	60–110	27–126	16–27	29–55
Mean	81.5	62.9	22.3	40.9
Standard dev.	13	26.7	3.9	11.9
Coeff. of var.	16	42	17	29
Mg				
Range	3–10	4–44	14–19	25–29
Mean	5.7	22.6	15.9	27.6
Standard dev.	1.4	6.3	2.0	2.2
Coeff. of var.	25	28	13	8
Na				
Range	1–16	3–95	73–136	130–369
Mean	5.2	30.6	93.3	256
Standard dev.	3.0	25.4	17.6	122
Coeff. of var.	58	83	19	48
Cl				
Range	1–38	2–42	23–50	175–490
Mean	7.9	26.9	39.7	351
Standard dev.	5.3	25.0	8.4	168
Coeff. of var.	67	93	21	8
SO$_4$				
Range	16–93	12–167	35–52	10–20
Mean	29.7	68.5	42.0	16.4
Standard dev.	12.2	42.8	8.0	4.5
Coeff. of var.	41	62	19	27
HCO$_3$				
Range	143–291	203–302	263–368	292–388
Mean	218	266	281	345
Standard dev.	40.2	25.7	31.3	54.7
Coeff. of var.	18	10	11	16
NO$_3$				
Range	0–55	0–24	0–2	0–2
Mean	30.2	3.5	0.16	0.45
Standard dev.	9.8	6.0	0.59	0.90
Coeff. of var.	32	171	369	200
Ca/Mg				
Range	6–16	0.7–4.9	0.7–1.1	0.7–1.2
Mean	8.9	1.9	0.86	0.90
Standard dev.	1.9	1.1	0.16	0.26
Coeff. of var.	21	58	19	29
TDS				
Range	161–135	211–513	308–437	536–1129
Mean	236	343	351	862
Standard dev.	43	74	34	310
Coeff. of var.	18	22	10	36

Table 13.2. Chemical variability as shown by coefficients of variation of TDS for wells (w) and springs (s) sampled at least six times (Ca subtype water is characterized by high chemical variability relative to Ca-Mg subtype water) (Scanlon 1990) (Reprinted with kind permission of Elsevier Science-NL, Sara Burgerhartstraat 25, 1055 KV Amsterdam, The Netherlands. Copyright 1990. All rights reserved)

Ca subtype		Ca-Mg subtype		Mixed water types	
Well/spring no.	CV TDS no.	Well/spring no.	CV TDS	Well/spring	CV TDS
4s	8.1	5w	3.0	32w	7.6
8s	9.7	7w	4.0	33w	6.0
26s	15.0	25w	3.0		
31s	13.9	37w	6.5		
48s	13.6	46w	2.6		
93s	12.0	47w	6.0		
6w	11.8				
35w	10.2				
92w	9.9				

13.3
Summary

The coefficient of variation can be used to evaluate the data in terms of dispersion characteristics. It is quickly and easily done on data to derive simple descriptive aspects.

13.4
Supplemental Reading

Griffiths JC (1967) Scientific methods in analysis of sediments. McGraw-Hill, New York

CHAPTER 14

Correspondence Analysis

14.1
Procedure

Correspondence analysis is defined in some instances as a way of interpreting contingency tables that may be defined through principal components analysis (Mardia et al. 1979). In correspondence analysis used in this book, a factor is represented by the eigenvector of the normalized covariance or correlation matrix (Usunoff and Guzman-Guzman 1989). It can be further viewed as a simultaneous linear regression scheme with dual scaling, which allows the interpretation of both sample sites and variables in the same factor space. The points, i (sample) and the points j (variables) can be simultaneously reported on the planes associated with the factor axes. The proximity of a point j to a group of points i is taken as an indication that the variable actually characterizes this group of samples. The contribution, CR of the points i or j, in the variability accounted for by an axis (a) can be computed and this aids interpretation. The CR of whole points i or j amounts to one, according to calculations (Razack and Dazy 1990):

$$\sum_i CR_\alpha(i) = \sum_j CR_\alpha(j) = 1 \tag{14.1}$$

for a given axis (α), and the quality (QT) which expresses the quality of representation of points on the axis (α) is given by:

$$\sum_\alpha QT_{(i)} = \sum_\alpha QT_{(j)} = 1 \text{ for a given point i or j.} \tag{14.2}$$

For a fuller treatment of the subject the reader is referred to Mardia et al. (1979). The parsimonious results that derive from this analysis are its strong points. The patterns in the data are easily determined through observation of the data plots.

14.2
Numerical Example on Ground-Water Mixing Zones

A correspondence analysis was applied to study the geochemistry of ground-water mixing zones in the French Alps, and conclusions are given below. Razack and Dazy (1990) used correspondence analysis to look at ground-water mixing zones in sedimentary and metamorphic aquifers in the western French Alps.

Values of physical and chemical parameters were determined on 95 water samples that related to Allevard spring. Their procedure was used to interpret hydrochemical data based on a combined use of correspondence factor analysis and Piper's classification. The approach proved successful in differentiating groundwater masses of different hydrological origin which mix and discharge at the spring outlet of the system. The correlations, principal components, and correspondence factor analytic results are shown in Tables 14.1, 14.2, 14.3, and 14.4. Figure 14.1 is a plot of (A) correlations between original variables and principal components, and (B) a plot of water samples on the first two principal components. Figure 14.2 is a graph of the correspondence factor analysis of raw water quality data. Figure 14.3 is a plot of constituent modalities of the variables temperature, H_2S, and SiO_2 on the plane associated with the first two axes, and a plot of the water samples.

Table 14.1. Matrix of correlation coefficients for chemical variables. (After Razack and Dazy 1990) (Reprinted with kind permission of Elsevier Science-NL, Sara Burgerhartstraat 25, 1055 KV Amsterdam, The Netherlands. Copyright 1990)

	Ca^{2+}	Mg^{2+}	Na^+	K^+	HCO_3^-	Cl^-	SO_4^-
Ca^{2+}	1						
Mg^{2+}	0.559	1					
Na^+	0.727	0.783	1				
K^+	0.265	0.182	0.206	1			
HCO_3^-	0.619	0.775	0.832	0.087	1		
Cl^-	0.735	0.771	0.947	0.141	0.811	1	
SO_4^-	0.866	0.727	0.900	0.270	0.719	0.822	1

Table 14.2. Principal components analysis of raw water quality data; correlation between original variables and the principal component-PC. (Razack and Dazy 1990) (Reprinted with kind permission of Elsevier Science-NL, Sara Bugerhartstraat 25, 1055 KV Amsterdam, The Netherlands. Copyright 1990. All rights reserved)

	Ca^{2+}	Mg^{2+}	Na^+	K^+	HCO_3^-	Cl^-	SO_4^{-2}
PC1	0.376	0.383	0.433	0.118	0.393	0.423	0.421
PC2	0.143	−0.095	−0.061	0.948	−0.212	−0.126	0.080
CR-PC1	0.835	0.851	0.962	0.262	0.873	0.940	0.935
CR-PC2	0.14	−0.09	−0.06	0.95	−0.212	−0.125	0.080

PC1, PC2 = coefficients of principal components 1 and 2 (coordinates of eigenvectors of the correlation matrix in Table 14.1); CR-PC1, CR-PC2 = correlation between the initial variables and PC1 and PC2.

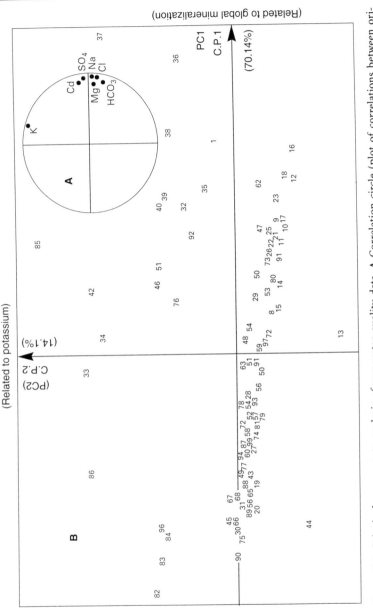

Figure 14.1. Principal components analysis of raw water quality data. **A** Correlation circle (plot of correlations between original variables and principal components; **B** plot of water samples on the plane associated with the first two principal components (After Razack and Dazy 1990) (Reprinted from Hydrology with kind permission of Elsevier Science-NL, Sara Burgerhartstraat 25, 1055 Amsterdam, The Netherlands. Copyright 1990. All rights reserved)

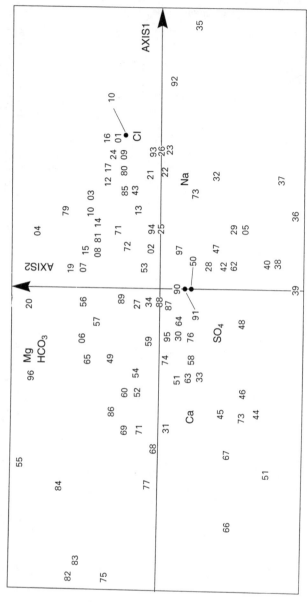

Figure 14.2. Correspondence factor analysis of raw water quality data. Simultaneous representation of the chemical constituents and of the water samples on the plane associated with the first two axes (Razack and Dazy 1990) (Reprinted from Hydrology with kind permission of Elsevier Science-NL, Sara Burgerhartstraat 25, 1055 KV Amsterdam, The Netherlands. Copyright 1990)

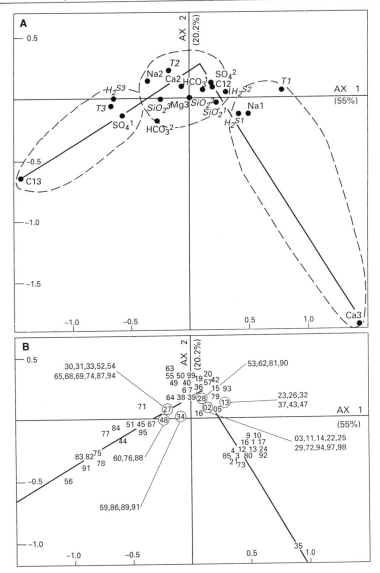

Figure 14.3. Correspondence factor analysis of the transformed data. **A** Plot of the constituent modalities and of the supplementary variables (T, H$_2$S, SiO$_2$) on the plane associated with the first two axes. **B** Plot of water samples. (Razack and Dazy 1990) (Reprinted from Hydrology with kind permission of Elsevier Science-NL, Sara Burgerhartstraat 25, 1055 KV Amsterdam, The Netherlands. Copyright 1990. All rights reserved)

Table 14.3. Correspondence factor analysis of the transformed hydrochemical data; contributions of modalities indicate roles of variables in defining factors 1 and 2. (After Razack and Dazy 1990) (Reprinted with kind permission of Elsevier Science-NL, Sara Burgerhartstraat 25, 1055 KV Amsterdam, The Netherlands. Copyright 1990. All rights reserved)

	Ca^2	Ca^3	Mg^3	Na1	Na2	HCO_3^2	HCO_3^3	Cl2	Cl3	SO_4^1	SO_4^2
A1	0.6	12.7	0	14.1	11.9	2.4	0.7	5.7	36.7	10.6	4.5
A2	2.9	59.3	0	3.9	3.3	2.6	0.7	3.1	19.8	3.1	1.3
Q1	33.7	33.7	0	6.7	6.7	18.8	18.8	79.0	79.0	46.0	46.0
Q2	57.5	57.5	0	6.7	6.7	7.3	7.3	15.7	15.7	4.9	4.9
Q3	91.2	91.2	0	73.3	73.3	26.1	26.1	94.7	94.7	50.9	50.9

A1, Contribution, CR, accounted for by axis 1;
A2, contribution, CR, accounted for by axis 2;
Q1, quality, QT, on axis 1 = quality of representation on axis;
Q2, quality, QT, on axis 2 = quality of representation on axis;
Q3, quality, QT, on axes 1 and 2 = quality of representation on axes. See also Eqs. (14.1) and (14.2).

Table 14.4. Correspondence factor analysis of raw chemical data. Contribution (CR%) and quality (QT%) of the initial variables with respect to axes 1 and 2. (Modified from Razack and Dazy 1990) (Reprinted with kind permission of Elsevier Science-NL, Sara Burgerhartstraat 25, 1055 KV Amsterdam, The Netherlands. Copyright 1990)

	Ca^{2+}	Mg^{2+}	Na^+	HCO_3^-	Cl^-	SO_4^{-2}
CR-axis 1	26.1	4.2	26.6	6.1	30.0	6.9
CR-axis 2	6.0	36.4	2.3	30.1	2.6	22.6
QT-axis 1	70.5	16.9	80.5	25.3	73.4	34.0
QT-axis 2	6.2	55.4	2.7	48.1	2.4	42.3
QT-axis 1–2	76.7	72.3	83.2	73.4	75.8	76.3

CR, Contribution to axis;
QT, Quality of representation on axis.

14.3
Numerical Example on Water Chemistry

The chemistry of waters from aquifers in Arizona were investigated with a correspondence analysis on the data. Usunoff and Guzman-Guzman (1989) applied factor and correspondence analysis to chemical data from two aquifers to assess their usefulness in deciphering waters from different aquifers in the San Pedro River basin, Arizona. They found that in using factor analysis they were able to extract two factors that were used to segregate waters. The correspondence analysis was applied only to the confined aquifer and two factors were found which related to dissolution of gypsum coupled with ion exchange as the main mechanisms operating in the aquifer, and presence of fluorite. The results of their study are shown in Figs. 14.4, 14.5, and 14.6.

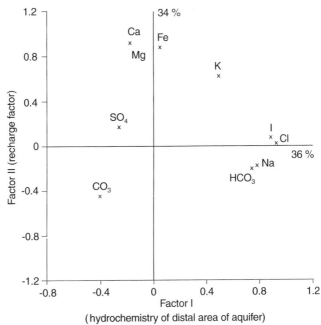

Figure 14.4. Milk River aquifer: distribution of variables as given by factor analysis (R-mode). (Usunoff and Guzman-Guzman (1989) (Reprinted by permission of Groundwater. Copyright 1989. All rights reserved)

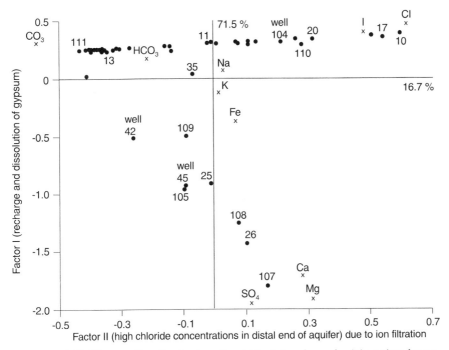

Figure 14.5. Milk River aquifer: distribution of variables (×) and samples (●) as given by correspondence analysis. (Usunoff and Guzman-Guzman 1989) (Reprinted by permission of Groundwater. Copyright 1989. All rights reserved)

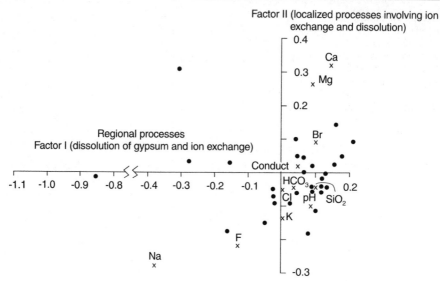

Figure 14.6. San Pedro aquifer: distribution of variables (×) and samples (•) as given by correspondence analysis. (Usunoff and Guzman-Guzman 1989) (Reprinted by permission of Groundwater. Copyright 1989. All rights reserved)

14.4
Summary

The simultaneous analysis of variables and samples is accomplished by using the corresponding analysis technique. It easily shows the relationships between variables and samples on factors.

14.5
Supplemental Reading

Davis JC (1986) Statistics and data analysis in geology. John Wiley, New York, 646 pp

CHAPTER 15

Multivariate Probit Analysis

15.1
Concept and Procedure

Probit analysis is used in the environmental toxicology field as a procedure to study the dosage response relation in a population of biological organisms, where randomly chosen population members are exposed to various levels of applied stimulus and quantal response is assessed as either dead or alive. In some instances more than one organ or physiological system is affected by the stimulus leading to tests of so-called main effects and side effects.

15.2
Numerical Example on Biological Data

A multivariate probit analysis was done on dosage response data. Ashford and Sowden (1970) discussed the analysis appropiate for a situation in which the samples from a population of biological organisms (in this case humans) are exposed to a stimulus at various levels, the effect on the organisms being manifested in terms of the observable quantal responses of two or more separate systems. Figure 15.1 and Table 15.1 define the final probit model parameters and chi-square test results. The actual study was designed to see if respiratory symptoms from coal mining work were manifested according to age of mine workers. The number of samples was 18000. Thus, the presence of a particular symptom may be taken as equivalent to the quantal response of the corresponding physiological system to a stimulus. Age in this study was taken as a measure of the coal mining effects. Figure 15.1 is a plot of age versus prevalence of respiratory symptoms. The scale used for prevalence was a nonlinear probit scale on standard probability graph paper.

15.3
Numerical Example on Agricultural Data

The second example of multivariate probit analysis considers the relation ship between chemical determinations of phosphorous in soil and plant-available phosphorous in corn grown in that soil. Table 15.2 shows the raw data.

Figure 15.1. Plot of age versus prevalence of symptoms. (Ashford and Sowden 1970)

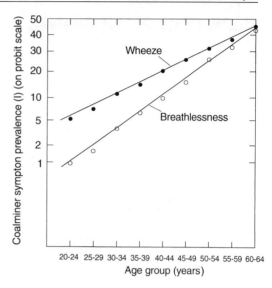

$a'_1 = -3.6113$	$b'_1 = 0.05484$
$a''_2 = -2.4542$	$b''_2 = 0.03700$
$p'_{12} = 0.7709$	

Chi-square = 20.8 with 22, $p = 0.5$

a'_1 and b'_1, are initial straight line parameter estimates.
a''_2 and b''_2 are final linear model parameter estimates.
p'_{12} is the correlation for age groups b and on initial and final parameter estimates.

Explanation:

Initial estimated effects functions for breathlessness is:

$x_1(z) = -3.60 + 0.055z$, where z = age

Initial estimated effects function for wheeze is:

$x_2(z) = -2.45 + 0.0372$

Final estimated effects function for breathlessness is:

$x_1(z) = 3.6113 + 0.05484\, z$

Final estimated effects function for wheeze is:

$x_2(z) = -2.4542 + 0.037002\, z$

Correlation between estimates is : 0.7709

The results of simultaneously testing for the effects of X1 and X2 with the probit model and multiple regression model are shown in Table 15.3. It also gives a comparison of multiple regression results and multivariate probit analysis. The modified procedure is based on the same distribution as multiple regression, and the significance levels agree closely with these two procedures. For any test, however, the conclusion is that inorganic phosphorus in the soil appears to be

Table 15.2. Inorganic phosphorous X1, a component of organic phosphorus X2, and plant-available phosphorus Y, in parts per million in 18 Iowa soils at 20 °C; and ordered categories of Y (Y') (From Snappinn and Small 1986. Reprinted by permission of Biometrics. Copyright 1986)

Soil sample	X1	X2	Y	Y'
1	0.4	53	64	3
2	0.4	23	60	4
3	3.1	19	71	3
4	0.6	34	61	4
5	4.7	24	54	4
6	1.7	65	77	3
7	9.4	44	81	2
8	10.1	31	93	2
9	11.6	29	93	2
10	12.6	58	51	4
11	10.9	37	76	3
12	23.1	46	96	1
13	23.1	50	77	3
14	21.6	44	93	2
15	23.1	56	95	1
16	1.9	36	54	4
17	26.8	58	168	1
18	29.9	51	99	1

Table 15.3. Tests of significance of the explanatory variables. (After Snapinn and Small 1986) (Reprinted by permission of Biometrics. Copyright 1986)

Source of variation	Ordinal test						Multiple regression		
	Standard test (Chi-square dist.)			Modified test (F-distribution)					
	G^2	df	P	G^2/q	df	P	F	df	P
X1 and X2	15.62	2	0.0004	7.81	2,15	0.005	6.99	2,15	0.007
X2 after X1	0.10	1	0.749	0.10	1,15	0.753	0.04	1,15	0.844
X1 after X2	12.63	1	0.0004	12.63	1,15	0.003	10.30	1,15	0.006

strongly related to plant-available phosphorus, but organic phosphorus does not (Snapinn and Small 1986). The studies here are in addition to those by Snedecor and Cochran (1967) on the same data.

15.4
Summary

Multivariate probit methods are used on data in a similiar fashion to multiple regression to determine variable relationships.

15.5
Supplemental Reading

Agresti A (1990) Categorical data analysis. John Wiley, New York

Multivariate Measures of Space, Distance, and Time

CHAPTER 16

Multivariate Time Series Modeling

16.1
Concept and Introduction

Time series models have been applied to many environmental and geohydrological problems. In many instances, such models may be required to provide the most accurate forecasts possible. Before proceeding, a short review of methods will be given.

Classical time series methods fall into three main areas: analysis, forecasting, and control. The analysis of many time series models has been presented by Box and Jenkins (1970). These techniques are very commonly used approaches in advanced forecasting and modeling. The Box-Jenkins approach has four main components: data transformation, model identification, parameter estimation, and diagnostic checking.

In general, the analysis of time series usually includes four related stages of development. The first stage involves data gathering. For example, in geohydrologic cases, data on the hydrologic system of interest are collected to determine how the system works, i.e., its structure. The second stage is concerned with system identification and specification, i.e., a determination of the elements constituting the ground water system, or surface water system is done. Stage three is involved with quantifying the relationships and with parameter estimation. The fourth stage, diagnostic checking, is used to check the model for goodness of fit to the data and its usefulness derives from providing a description of the system. In summary, a forecasting system is more than a technique, because it includes data collection, data management, methodology, experimentation, incorporation of judgemental data, and continuous monitoring for effectiveness of parts thereof. Table 16.1 provides a classification of several methods applied in the sciences.

16.2
Definitions

Times Series. Mathematically, a time series is defined by the values, $Y(1)$, $Y(2) \ldots Y(p)$ of a variable Y at times $t(1), t(2) \ldots t(n)$. Thus Y is a function of time. Graphic analyses of data are important first steps in any time series analysis.

Secular Variation. The secular variation or secular trend is the long term movements or component referred to on a graph that depicts where a time series is

Table 16.1. Some commonly used forecasting methods in time series analysis (from simplest to more complex)

Method	Some examples or characteristics
1. Judgmental	Simplest, moving average, most recent value type
2. Deterministic	Curve extrapolation, polynomial, growth, etc.
3. Exponential smoothing	Simple exponential, Holt two parameter, Winters three parameter (nonlinear)
4. Decomposition	Classical, ARIMA, census X11
5. Bayesian forecasting	Somewhat obsolete
6. Box-Jenkins (ARIMA) models	Basically univariate, robust, multivariate, transfer functions
7. State Space Forecasting	Univariate, multivariate, similiar to Box-Jenkins approach, uses transfer functions, uses canonical correlation
8. Dynamic regression	Assumes regression model, causality model
9. Variable parameter regression	
10. Bayesian vector autoregression	Multivariate, regression of endogenous variables on lagged (driving) endogenous and exogenous (external) variables

going over some long period of time. It is defined by a trend line and must usually be determined for treatment. The trend component is the component that increases or decreases smoothly with time. This trend can be determined by using moving average methods, regression, exponential smoothing, or similar.

Cyclical Variations. Cyclical movements or cyclical variations, which may or may not be periodic, refer to the long-term oscillations or swings about a trend line or curve. These cycles as they are called, are usually denoted only if they occur or recur after time intervals of greater than 1 year and must be treated properly.

Seasonal Variations. Seasonal movements or seasonal variations refer to the identical, or almost identical patterns which a time series appears to follow during corresponding months of successive years (i. e., annual events). This concept is usually referred to as annual periodicity but it may be extended to include daily, hourly, weekly, or other time periods. It can be analyzed by such methods as a ratio to moving average method (Downing and Clark 1985). When the cyclical fluctuations and the trend have been subtracted from the data, the random fluctuations that are left are referred to as random error or residuals.

Stationary Process. The average of a process is not changing over time.

Differencing. A method for reducing a nonstationary process to one that is stationary.

Scatter Plot. A plot of the time series data to evaluate if the data is approximately stationary.

Correlogram. A tool for assessing the time-lagged correlations (autocorrelations) of a time series. It is a graph of the autocorrelations.

16.3
Treatment of Time Series Data

Time series data is effectively operated on by several means using many software packages such as Minitab, SAS, or others.

1. The first operation is deseasonalization of the data, where if monthly, the data are divided by the corresponding seasonal index numbers, the result of which is said to be deseasonalized or adjusted data for seasonal variation. These data still include trend, cyclical, and irregular movements which must somehow be accounted for.
2. The second operation is estimation of cyclical variations, whereby, after data is deseasonalized, they can be adjusted for trend by simply dividing the data by corresponding trend values.
3. The third important operation is estimation of irregular or random variations. The estimation of irregular or random variations is achieved by adjusting the data for trend, seasonal, and cyclical variations.
4. The fourth operation involves estimating the comparability of data, i.e., in case the measurement period is variable, as with monthly data (28 to 31 days).
5. The vital fifth operation is forecasting, or extrapolating to some future time, and can be done by a number of methods such as exponential smoothing or other techniques.

Some aspects of time series analysis have been presented by Fox (1975). Several techniques for smoothing time series included in this discussion are simple moving average, polynomial curve fitting, iterated moving averages, and Fourier analysis. Fourier analysis uses a series of sine and cosine curves representing fundamental harmonics to fit the data. Polynomial curve fitting and iterated moving averages make use of summation equations that are easily handled by computers. The reader is referred to Fox (1975) for examples of data analysis that involve simple time series.

16.4
Multivariate Time Series Procedures

16.4.1
Types of Multivariate Models

The analysis of multivariate time series modelling will not be treated in great detail as only a general introduction is possible in this text. The reader is referred to several other papers which are comprehensive in scope and treatment of the subject (Lungu and Sefe 1991; Frost and Clarke 1973; Jenkins 1982; Makridakis et al. 1982; Gardner 1985; Makridakis 1986; Chow and Kareliotis 1970). The family of methods that are commonly used to describe multivariate time series include: (1) dynamic regression methods that link time series and causal affects; (2) Box-Jenkins autoregressive integrated moving average (ARIMA) methods that use rational polynomials for parsimonious representation of the autocorrelation function of a time series; and (3) state space methods that are statistically the same or similiar to Box-Jenkins (ARIMA) but are truly multivariate and use canonical correlation theory to determine the mode and identify model parsimoniously.

The above methods usually employ some type of "best fit" information criterion such as: (1) Aikaike AIC (Akaike information criterion = $S^2 \exp(2n/T)$, where S^2 is the mean square residual error, n is the number of fitted parameters, and T is the number of points in sample space), which balances goodness of fit to the historical data and model complexity and is used across methodologies to compare models, or (2) Schwarz BIC ($S^2 T^{n/T}$) that is similiar to AIC but penalizes complexity more severely in model selection. For other discussions of concepts in time series analysis, many textbooks exist and the reader is referred to the group of publications referenced in this text.

16.4.2
Box-Jenkins (ARIMA) Method

Box-Jenkins methods are correlational techniques that use both the autoregressive and moving average components, and this aspect implies that correlations in the past are used to forecast into the future. This extrapolation is not valid if data are statistically unstable (nonstationary, i.e., the structure of data changes significantly in time) as described below. If data are homogeneous, however, and stable, then the Box-Jenkins method is likely to outperform other techniques such as exponential smoothing in forecasting.

The Box-Jenkins ARIMA method is used for univariate, multivariate, and transfer function analysis of time series data. It uses rational polynomials for parsimonious representation of the autocorrelation function of a time series as follows (Goodrich and Stellwagen 1991):

$$y_t = P_z/Q_z e_t \tag{16.1}$$

where P_z is the p'th order polynomial, Q_z is the q'th order polynomial, and e_t = lag errors.

Often a nonstationary time series can be transformed by differencing or detrending to obtain stationarity. For a stationary process, the autocorrelation function (ACF) decays with lag m, i.e., it dies or decays out quickly. For large data sets, we will most often find the data to be nonstationary. If the time series is stationary, then the expected value of the sample mean is independent of time. For a stationary process, the properties are unaffected by choice of time origin, or starting point. For a stationary process, the diagonals are constant for the covariance matrix.

The ARIMA model is a good performer for short time horizons, is a powerful method to analyze statistical interdependency of a series, and is a probability model. The time series must be preprocessed to stationarity, and both multivariate and transfer function models tend to be difficult to fit and explain. These methods, however, behave very well when normality and stationarity exist in the data.

16.5
Numerical Example on Air Pollution Data

Time series analysis using Box-Jenkins methods were applied in the study of sulfur dioxide concentrations. Hsu and Hunter (1975) analyzed air pollution concentrations with seasonal variations using annual time series records of daily averages of hourly sulfur dioxide concentrations recorded over four major cities: Chicago, Philadelphia, St. Louis, and Washington, D.C. All the series exhibited strong seasonal patterns in both the level and variance, and to construct models appropriate and useful for prediction and analysis, a Box-Cox transformation was used to stabilize the data variability. The transformed data were then modeled using simple seasonal plus stochastic components following Box-Jenkins time series methodology. The transformed data was then fitted by a cosine curve to model the seasonal influences and, following the techniques suggested by Box and Jenkins (1970), a stochastic model fitted to account for the day-to-day time-dependent character of the data. Based on autocorrelations for the data, a moving average model of the first order was chosen for analysis because the first autocorrelation value was significantly larger than zero at the 0.05 significance level. The residuals were investigated using the sample lagged autocorrelation coefficients. The fitted models, forecasts, and confidence limits are shown in Figs. 16.1, 16.2, and 16.3. Parameter estimates were derived for the residuals from the time series model of daily averages of SO_2 concentrations collected from the major cities investigated. The analysis done in this study shows the complex nature of time series modeling, especially for the multivariate case.

The reader is referred to Hsu and Hunter for a complete summary of the model parameters and more detail.

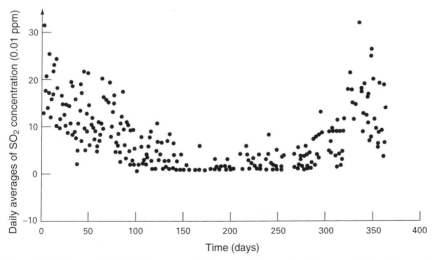

Figure 16.1. Observed daily SO_2 concentrations in Chicago for the year 1969. (Hsu and Hunter 1975)

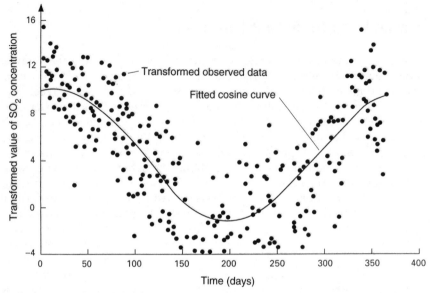

Figure 16.2. Transformed daily SO_2 concentrations in Chicago for the year 1969 and a fitted cosine curve. (Hsu and Hunter 1975)

Figure 16.3. Forecasting SO_2 concentrations in Chicago for the first 100 days of 1970. (Hsu and Hunter 1975)

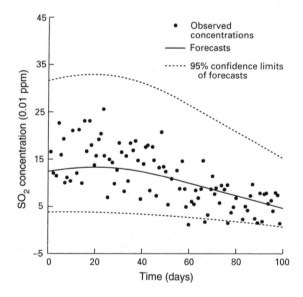

16.6
State Space Models

These models are subsets of formerly discussed models and assume that the parameters in the model change each time an observation comes to hand. State space modeling can handle multiple endogenous and exogenous variables. State space methods are statistically very similar to Box-Jenkins methods, but are truly multivariate in nature. They are used in various formats for univariate, multivariate, and transfer function models. A 'markov' property implies the process wherein the state of the position of the property on which it depends is only the last position. The covariance matrix of a stationary system (process) can be realized by a linear Markov model in the state space form of the equation below. One method uses canonical correlation theory to identify the model and uses a two equation matrix recursion to represent dynamics in the following form (Goodrich and Stellwagen 1991):

$$x(t + 1) = F\,x(t) + G\,z(t + 1) + K\,v(t)$$
$$y(t) = H\,x(t) + v(t), \tag{16.2}$$

where: $x(t)$ is the px1 state vector; p is the order of polynomial; $z(t)$ is a qx1 exogenous vector; $v(t)$ is a process error vector; $y(t)$ is a rx1 endogenous vector; F is the pxp transition matrix; G is the pxq exogenous effect matrix; and H is the rxp measurement matrix.

The state vector $x(t)$ consists of relevant information from the past. This method requires large amounts of data to work most efficiently, i.e., 40 to 50 data points. The time series must also be processed to stationarity. The multivariate time series in this case are as easy to analyze as they are for the univariate data. State space models behave very well when normality and stationarity are exhibited by data. Much of the effort in state space modeling is to predict the covariances which are used for the canonical correlation technique.

16.7
Markov Process Models

Markov analysis is used to study the behavior of a variable through time. Markov models have several inherent properties, and they are used to analyze a process which exhibits the following properties. First and foremost, there must be a finite set of possible outcomes. The second required property of the process is that the probability of the next outcome, called a transition probability, is entirely dependent on the prior outcome only. The third required property of the process is that the probabilities remain constant over time.

16.8
Dynamic Regression Models

Dynamic regression models involve the integration of regression and time series analysis. It is accomplished by adding a lag variable to regression. The dynamic regression model combines traditional ordinary least squares (OLS) with a time series approach. Computationally, the model is estimated as though the lagged dependent variables were independent lagged variables. Practically speaking, we can ignore the difference between lagged dependent variables in the process of model construction.

The basic OLS dynamic regression model equation is given by the following equation where (Goodrich and Stellwagen 1991):

$$P(z)\, y(t) = B\, X[t] + e(t), \tag{16.3}$$

$e(t)$ is independent and identically distributed, $N(0,var)$, $t = 1, 2, \ldots T$; $y[t]$ is the dependent variable at time t; $X[t]$ is a n-vector of exogenous variables; $P(z)$ is a polynomial in the backwards operator z; B is a row n-vector of coefficients; and $e[t]$ is the residual error at time t.

The long version of equation (16.3) above is derived as follows:

$$P(z) = 1 - az - a(z^2) - \ldots - a\, z^p, \tag{16.4}$$

and

$$y[t] = a(1)\, y[t-1] + a(2)\, y[t-2] + \ldots + a(p)\, y[t-p] + B\, X[t] + e[t], \tag{16.5}$$

with definitions as above.

Some of the major problems in dynamic regression include model specification (i.e., how many parameters should be incorporated), but in some cases the most common and damaging problem is serial correlation which we detect by examining the residual autocorrelation function (ACF), using the Durbin-Watson statistic (Durbin and Watson 1950, 1951, 1971), or Ljung-Box (Ljung and Box 1978) statistic.

16.9
Multivariate Stochastic Models

These models are generalizations of univariate models and the goal of the models are to explain mutual interaction or feedback between more than one time series. Applications in analyses of commodities have receive major benefits, but applications in geohydrology (Chow and Kareliotis 1970; Lungu and Sefe 1991) are less well known.

16.10
Numerical Example on Stochastic Analysis of Hydrologic Data

The watershed chosen for study is located in the upper Sangamon river basin, is 550 sq. miles in areal extent, and is situated above Monticello in central Illinois. Chow and Kareliotis (1970) treated this watershed as a stochastic hydrologic system whose components included precipitation, runoff, storage, and evapotranspiration, which are simulated as stochastic processes by time series models used to determine correlograms and spectral patterns (spectral analysis). They used an autoregression model, correlogram, and spectral analysis to determine the best and most appropriate model. In this study, they treated the stochastic processes of precipitation, conceptual water storage, and evapotranspiration not as independent components but as a three-dimensional vector or multiple time series. The assumptions were: (1) each stochastic process consists of two parts, one deterministic and the other random and uncorrelated to the deterministic part and the parts of other processes, and (2) the deterministic part of each stochastic process consists also of two parts, one depending only on time and the other depending on the vector of the stochastic processes of precipitation, conceptual watershed storage, and actual evapotranspiration at previous time intervals (Chow and Kareliotis 1970). They determined mass curves and correlograms for precipitation, conceptual watershed storage, and evapotranspiration, and spectra of precipitation, conceptual water storage, and evapotranspiration and those results are presented (Figs. 16.4–16.7). Spectra for the variables were also done (Figs. 16.8–16.10). They found that the expected values of the system components of precipitation, conceptual water storage, and evapotranspiration were found to be simulated by harmonics of 12-month and 6-month periodicities (Chow and Kareliotis 1970).

Figure 16.4. Mass curves for hydrologic varables. (Chow and Kareliotis 1970) (Reprinted with permission of Water Resources Research. Copyright 1970. All rights reserved)

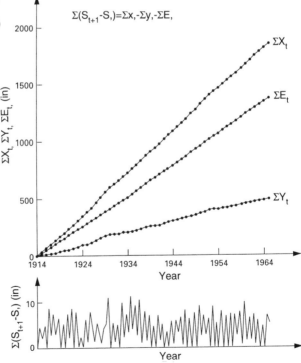

Mass curves of precipitation, evapotranspiration, runoff, and conceptual watershed storage. $(\Sigma(S_{t+1}-S_t)=\Sigma X_t-\Sigma Y_t-\Sigma E_t)$.

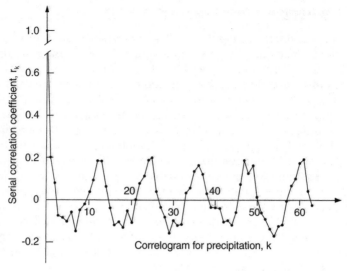

Figure 16.5. Correlogram for precipitation. (Chow and Kareliotis 1970) (Reprinted with permission of Water Resources Research. Copyright 1970. All rights reserved)

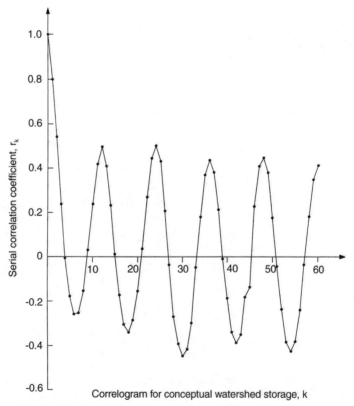

Figure 16.6. Correlogram for conceptual water storage. (Chow and Kareliotis 1970) (Reprinted with permission of Water Resources Research. Copyright 1970. All rights reserved)

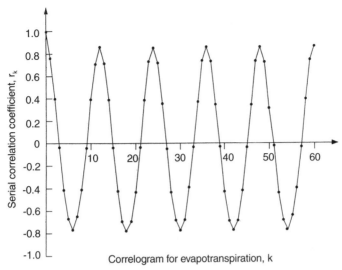

Figure 16.7. Correlogram for evapotranspiration. (Chow and Kareliotis 1970) (Reprinted with permission of Water Resources Research. Copyright 1970. All rights reserved)

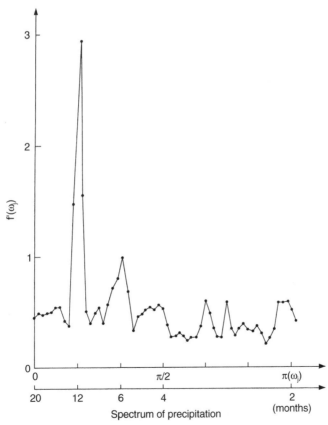

Figure 16.8. Spectrum of precipitation (Chow and Kareliotis 1970) (Reprinted with permission of Water Resources Research. Copyright 1970. All rights reserved)

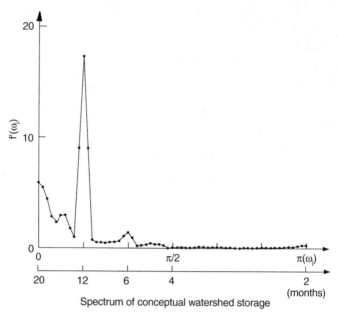

Figure 16.9. Spectrum of conceptual watershed storage. (Chow and Kareliotis 1970) (Reprinted with permission of Water Resources Research. Copyright 1970. All rights reserved)

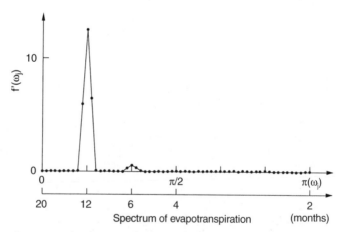

Figure 16.10. Spectrum of evapotranspiration. (Chow and Kareliotis 1970) (Reprinted with permission of Water Resources Research. Copyright 1970. All rights reserved)

16.11
Multivariate Transfer Functions

Transfer function models are made up of two components, a component (or components) exactly explained by the explanatory variable (or variables) called transfer functions or lag structures, and a component due to all the other variables not included in the model – referred to as noise or error structure.

Transfer function models aim to sort out how much weight in a forecast should be given to the past history of the series being forecast and how much weight should be given to the explanatory variables. This application is similar to the use of regression models that are unidirectional and assume that the explanatory variables Xi affect Y being forecast but no effect of Y on the Xi's exists.

Multivariate transfer function models are important tools because they can be used to investigate the relationship between several outputs (Yi) and several inputs (Xj).

16.12
Intervention Models

Intervention models are used to analyse the effects of major shocks that may give rise to large residuals in the model structure and thus distortions in the model parameter estimates and forecasts. These distortions may be removed by grafting dummy variables in the form of pulses and steps, each with its own transfer functions, into the time series models.

16.13
Diagnostics

Diagnostics are statistics which indicate whether there is a problem with the model. Different diagnostics are sensitive to different problems. The Durbin-Watson statistic is sensitive to first lag serial correlation. The Ljung-Box statistic is sensitive to overall correlations in the residuals. Diagnostics jeopardize the null hypothesis by postulating specific alternative hypotheses. They may be used to analyse time trends, excluded variables, dependent variable dynamics, serial correlation, nonlinearity, structural change, heteroscedasticity, and time-varying parameters.

16.14
Numerical Example on Ground-Water Data

Regression analysis and time series modeling were used to determine ground-water potential of a well field in Zambia. Houston (1983) applied time series analysis to the study of ground-water systems. He determined that water level fluctuations in a well field are dependent upon pumping rates and prior rainfall and can be simulated by a multiple linear regression model. The rate of dewatering of a mine is shown to be dependent upon antecedent mine size and rainfall, and can also be simulated by a multiple regression model. He concluded that such models can be used for forecasting and control of the ground-water systems, and once formulated are ideally suited to the management of ground-water systems where more costly and complex methods cannot be used.

Houston (1983) discussed the determination of reliable aquifer yields through three approaches: (1) a water balance equation, (2) an analytical approach such as Theis (1935), or (3) a numerical approach which simulates ground-water behavior based on various physical laws. The application of time series techniques to ground-water problems is a logical extension of the water balance equation approach, allowing a water balance equation to become dynamic and thus simulate a ground-water system in time (Houston 1983). A systems model essentially consists of input which is acted upon by a transfer function in order to produce output. The input data may be represented by recharge and discharge and the output by water level data. The transfer function of a systems model is estimated a posteriori by statistical techniques which produce a minimum error. In this way, the system's model differs fundamentally from analytical or numerical models which are based on physical laws established a priori, but is not assumed to replace or substitute for them. However, it can be used successfully where there are insufficient data to attempt other methods. It is also very cheap to develop, since pumping tests or other tests are not required to obtain estimates of permeability or storage. Furthermore, the simple mathematical expression which represents the model can usually be understood intuitively and can therefore be used to manage ground-water resources in situations where more complex methods would never be considered.

Houston (1983) derived the water balance equation with an autoregressive term to conform to structure of a ground-water system of interest. Cross-correlations between water level and rainfall were examined by using the cross-correlation function of two sequences. To look at fluctuations in water levels that were dependent upon antecedent water level, the autocorrelation function of a sequence was examined. Both cross-correlation and autocorrelation assume stationarity and linearity, and if these assumptions are not met, the mathematical basis is violated. The Broken Hill well field in Kabwe, Zambia was studied using the above methods. Ninteen boreholes in the well field penetrated a Precambrian dolomitic limestone aquifer of the Broken Hill series which forms a syncline to the northwest of Kabwe. A water balance equation for the Broken Hill area was derived as follows:

$$\Delta \text{ head} = \text{recharge factor} - \text{evapotranspiration} - \text{discharge(sw)} - \text{pumping(Q)} + \text{recharge(gw)}. \tag{16.6}$$

Figure 16.11 shows the correlograms of variables associated with the monthly data model of Broken Hill well field. Houston (1983) found that the autocorrelations of Δ head after detrending shows significant lags at 11 to 13 months, indicating an annual cycle as a result of its dependence on the recharge factor which also shows and clearly demonstrates an annual cycle in its autocorrelations. He also found that the cross-correlation of Δ head with recharge factor shows a signifi-

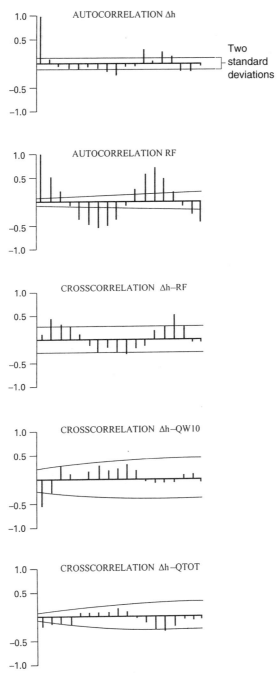

Figure 16.11. Correlograms of variables associated with the monthly data model of Broken Hill well field. (Houston 1983) (Reprinted by permission of Ground Water. Copyright 1983. All rights reserved)

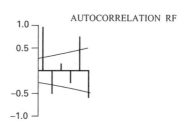

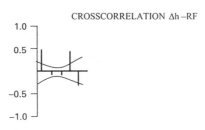

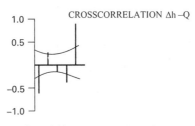

Figure 16.12. Correlograms of variables associated with annual data model of Broken Hill well field. (Houston 1983) (Reprinted by permission of Ground Water. Copyright 1983. All rights reserved)

cant positive lag at 1 month indicating the dependence of water level on prior rainfall. Additional significant lags were found to occur at 2 months due to further recharge attenuation and 14 months as a result of the annual cycle (Fig. 16.12).

16.15
Fit of Autoregressive Models

Given that serial correlation is a feature of a hydrological sequence, it has been customary to attempt a mathematical description of it, i.e., to fit a model by means of a low-order autoregression or, if data from two or more such sequences are available, by means of a bivariate or multivariate autoregression (Frost and Clarke 1973). A use for autoregression models in hydrology has been for the generation of synthetic sequences of hydrologic variables. The synthetic sequences have served two purposes: (1) investigating operating schemes of complex hydrologic systems with the objective of minimizing a given risk function; and

(2) estimating the frequency of occurrence of certain events, such as floods exceeding a given magnitude, of interest to the hydrologist. The second use exemplifies the Monte Carlo method for evaluating a multidimensional integral, although it is rarely stated to be such, and the first exemplifies the Monte Carlo method for seeking the minimum of a function too cumbersome for analytical treatment (Frost and Clarke 1973).

Frost and Clarke (1973) considered a problem that required the estimation of parameters characterizing a serially correlated hydrologic time series, yt, where data are also available from a longer time series, xt, itself serially correlated and cross correlated with yt. On the assumption that both series are lag one autore-gressions (when yt is characterized by three parameters μ, B, and var (t)), large sample variances for the autoregressive parameter B are derived from the short series yt alone or from both series xt and yt, and the relative information is investigated numerically. They concluded that when two serial correlations are approximately equal and the length of xt is twice that of yt, the gain in precision of the estimate of B is about 4% when the cross-correlation is 0.2, about 15% when the cross-correlation is 0.4, about 31% when the cross correlation is 0.6, and about 49% when the cross correlation is 0.8. The equations giving maximum likelihood (ML) estimates were also examined and a simple numerical technique was developed for one particular case of practical importance.

Other work (Carlson et al. 1970) suggests that autoregressive models or such generalizations as the ARIMA models may have use in hydrology, i. e., to provide forecasts of the magnitudes of future discharge events. These models have received use for forecasting in other related fields of science and have the advantage that approximate confidence limits for forecasts can be obtained easily. The mathematical treatment of the problem is given elsewhere and the reader is referred to this work for further insight (Carlson et al. 1970).

16.16
Summary

Time series modeling is a means of analyzing data wherein the dependent variable is time. Many different forms of models exist for geohydrologic application, including multivariate forms.

16.17
Supplemental Reading

Chatfield C (1984) The analysis of time series data – an introduction. Chapman and Hall, New York

CHAPTER 17

Multivariate Spatial Measures

17.1
Multidimensional Scaling

Multidimensional scaling is a technique for reducing dimensionality and is based on the distance between points. The result is a reduction in the dimensions used to represent the data with as little distortion as possible. It is similar in its objective to principal components analysis discussed earlier. If the methods employed use rank data, then it is called non-metric scaling otherwise, the terminology used is metric scaling. Studies concerning the analysis of distances between places on a map have received great attention in multidimensional scaling applications. The reader is referred to other references for additional insight (Chatfield and Collins 1980; Marriott 1974; and Mardia et al. 1979).

The technique of principal coordinate analysis under multidimensional scaling methods is related to principal components analysis because the principal coordinates of a data matrix [X] in k dimensions are given by the centroid scores of the n objects on the first k principal components (Mardia et al. 1979).

17.2
Regionalized Variables and Spatial Correlation

17.2.1
Introduction

The underlying mathematical basis for looking at spatial statistics in the geosciences derives from the theory of regionalized variables i.e., those variables that have some degree of spatial correlation (Matheron 1971). The basis of this theory is that a geologic process active in the formation of a mineral deposit or any other chemical or fluid body of matter, is interpreted as a random process. Examples of regionalized variables are grade of ore, thickness of a formation, or elevation of the surface of the earth, etc. Because the formation of the body is a random process, the grade or concentration at any point in a deposit or fluid body, is considered as a particular outcome of this random process. This probabilistic interpretation of a natural process as a random process is necessary to solve the practical problems encountered in estimating geologic or hydrologic variables. The method assumes that the adjoining samples are correlated to one another, which is directly opposite to classical statistics. It assumes that the particular relationship expressing the extent of the above correlation can be analytically and statistically captured in a function known as the variogram function or variogram in short form. The function $Z(x)$ displays a random aspect consisting of highly irregular and unpredictable variations, and a structured aspect reflecting the structural characteristics of the regionalized phenomena.

The analysis of spatial continuity and cross-continuity of variables such as thickness of a geologic horizon or concentration of a pollutant, soil strength measurements or permeabilities, rainfall measurements or diameter of trees all lend themselves to be analyzed with variograms, covariance functions, or correlograms. The analysis may occasionally not give a true description of the spatial continuity, however much can be learned from such failed attempts (Isaaks and Srivastava 1989). A complete summary of an analysis of an earth science dataset, the Walker Lake area of Nevada, is given in the text by Isaaks and Srivastava (1989) but the work will not be summarized here, because it is extensive. I will, however, present a few basic discussions about these topics in the sections below.

17.2.1
Concepts of Multivariate Geostatistics

The theory of regionalized variables has been used in a variety of scientific studies (Journel and Huijbregts 1978; Webster 1978; Davis 1986; Armstrong 1988; Ahmed and de Marsily 1989; Isaaks and Srivastava 1989; Smyth and Istok 1989; Webster and Oliver 1989). It provides a convenient way of summarizing spatial variability in the form of an auto semi-variogram which can be used to estimate weights for interpolating the value of a given property at an unsampled site. This technique is known as kriging and is a method of weighted local averaging that is optimal in the sense that it gives estimates of values at unrecorded sites without bias and with minimum known variance. As with most real life problems including kriging, the variables encountered are multidimensional. A logical extension to the situation when analysis is done where two or more variables are spatially interdependent and one is undersampled is a method called co-kriging. Co-kriging is useful for interpolating a property that is more difficult and costly to measure. Co-kriging makes use of the large spatial correlation between the property of interest and one of the less difficult to measure properties. The term "co-regionalization analysis" has been referred to as the analysis of co-regionalization matrices obtained from multivariate spatial data using a nested variogram model (Wackernagel et al. 1989). An example of application is given in Section 17.3. The application of canonical analysis, principal components analysis, and discriminant functions in co-regionalization analysis is a recognized way of dealing with co-regionalization matrices. Many other ways are possible for analyzing co-regionalization matrices because there are many ways to decompose a real symmetric positive semi-definite matrix. The reader is referred to Wackernagel et al. (1989) and Jolliffe (1986) for a more comprehensive treatment of this topic.

Co-regionalization analysis using principal components has been applied for geochemical data (Wackernagel 1988), pedological data (Wackernagel et al. 1988), ecology (Gittins 1985; Muge and Cabecadas 1989), and mining (Isaaks and Srivastava 1989) and the reader can gain more insight from those papers. The disjunctive co-kriging method that has been applied (Muge and Cabecadas 1989) and formulated by Matheron (1976), is also a multivariate technique for analyzing a variety of earth science data including pollution data.

17.3
Numerical Example on Soil Pollution Data

Co-kriging was applied to soil pollution data collected from the floodplain soils of the Geul River in Belgium and the Netherlands. The floodplain soils of the Geul River were studied to ascertain the co-regionalization between eleva-tion data and zinc concentrations in soils. The co-regionalization technique (co-kriging) was used to map the zinc concentrations from 154 observations (Leenaers et al. 1989).

The Geul River flows into the the Meuse River and is located in parts of Belgium and the southern Netherlands. Metal ore mining occurred extensively in the Geul Basin and mine tailings contributed vast amounts of unclaimed ore particles, such as lead, zinc, and cadmium to soils and river deposits.

A co-kriging procedure employing readily available elevation data in addi-tion to laboratory measurements of zinc concentrations, was used to construct a map of zinc concentrations in floodplain soils (Leenaers et al. 1989). The co-kriging procedure provided a map of zinc concentrations as well as a map of estimation variances. The maps were further compared with maps constructed from simple linear regression of zinc versus relative elevation and by point kriging. A subset of 45 samples from the original data consisting of 199 sam-pling sites was used. The 199 topsoil (0 – 10 cm; 100 g) samples were collected in a 5 km long length of the floodplain area of the Geul River, which has a width of 300 – 600 m.

One hundred and fifty-four of the 199 samples were used for computing the semivariograms for zinc and for cross-semivariogram between zinc and relative elevation and for point kriging of zinc. The other 45 samples were used for validating the interpolations. For co-kriging, the 154 data points were supple-mented by 309 spot heights; the semivariogram of relative elevation was also computed from these 463 relative elevation data points. Tables 17.1 and 17.2 show the descriptive statistics and correlation coefficients for the sediments in the Geul valley. Table 17.3 shows the parameters of the exponential semivario-gram models and Table 17.4 gives the summary statistics of absolute and squar-ed estimation errors. The interpolated maps are shown on Figs. 17.1 – 17.6.

From the analysis, it was clearly shown that use can be made of elevation data in a co-kriging procedure of mapping alluvial topsoils polluted with zinc in the floodplain of the Geul River. Despite the weak correlation between topsoil zinc and relative elevation it was possible to obtain better estimates of zinc concen-trations by co-kriging than by point kriging or simple linear regression. More-over, the variances of the estimates of co-kriging are substantially less than those obtained by point kriging (Leenaers et al. 1989).

Table 17.1. Measures of central tendency and variation of relative elevations (cm) and of zinc concentrations (mg/kg) in alluvial and colluvial sediments in the Geul valley. (After Leenaers et al. 1989) (Reprinted with kind permission of Kluwer Academic Publishers. Copyright 1989. All rights reserved)

	N	Mean	Median	Minimum	Maximum	Variance
Zinc content of alluvial sediments	154	741	543	114	2270	292877
Zinc content of colluvial sediments	12	147	124	68	338	5112
Relative elevation	463	429	411	224	791	9032

Fig. 17.1. Sample locations in part of the study area. (Leenaers et al. 1989) (Reprinted from Geostatistics, Vol. 1, pp. 371–382, with kind permission from Kluwer Academic Publishers. Copyright 1989. All rights reserved)

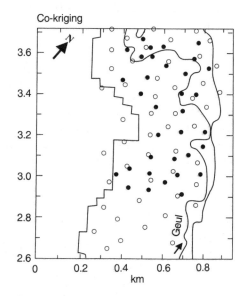

Co-kriging

Fig. 17.2. Auto semi variogram of zinc content in topsoil. (Leenaers et al. 1989) (Reprinted from Geostatistics, Vol. 1, pp. 371–382, with kind permission of Kluwer Academic Publishers. Copyright 1989. All rights reserved)

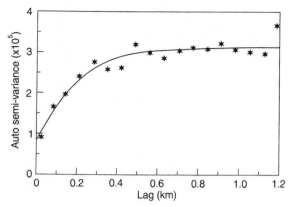

Table 17.2. Correlation coefficients of linear relations between zinc concentrations (Zn) and relative elevations (RE), n = 154. (After Leenaers et al. 1989) (Reprinted with kind permission of Kluwer Academic Publishers. Copyright 1989. All rights reserved)

	Zn	$10 \log(Zn)$
RE	−0.37	−0.48
$10 \log(RE)$	−0.36	−0.47

Table 17.3. Parameters of the exponential semivariogram models. (After Leenaers et al. 1989) (Reprinted with kind permission of Kluwer Academic Publishers. Copyright 1989. All rights reserved)

	Nugget (CO)	Sill (CO+C)	Range (a')
Auto semivariogram Zn	79262	314958	0.59
Auto semivariogram RE	849	9564	0.38
Cross semivariogram Zn–RE	−4254	−20351	0.73

Fig. 17.3. a Auto semi vario-gram of relative elevation and **b** cross semivariogram of relative elevation and topsoil zinc content. (Leenaers et al. 1989) (Reprinted from Geostatistics, Vol. 1, pp. 371–382, with kind permission of Kluwer Academic Publishers. Copyright 1989. All rights reserved)

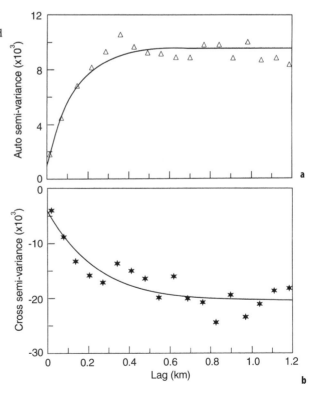

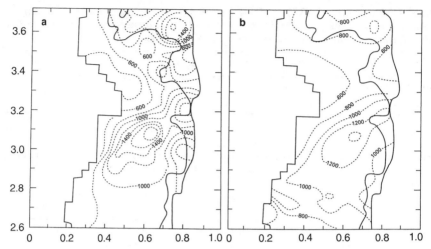

Fig. 17.4. Interpolated maps of top soil zinc content: **a** by point kriging; and **b** point co-kriging. (Leenaers et al. 1989) (Reprinted from Geostatistics, Vol. 1, pp. 371–382, with kind permission of Kluwer Academic Publishers. Copyright 1989. All rights reserved)

Fig. 17.5. Estimation variance of zinc: **a** point kriging; and **b** point co-kriging. (Leenaers et al. 1989) (Reprinted from Geostatistics, Vol. 1, pp. 371–382, with kind permission of Kluwer Academic Publishers. Copyright 1989. All rights reserved)

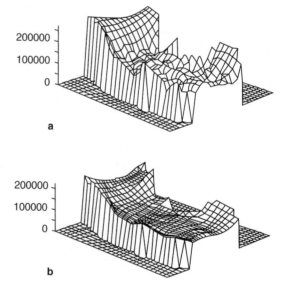

Table 17.4. Summary statistics of absolute and squared estimation errors (n = 45). (After Leenaers et al. 1989) (Reprinted with kind permission of Kluwer Academic Publishers. Copyright 1989. All rights reserved)

		Mean	Standard deviation	Median	Range	Skewness
Regression	(AE)	471	391	353	1412	0.91
	(SE)	371772	552497	124357	1993381	1.34
Kriging	(AE)	380	327	311	1397	0.63
	(SE)	249221	393494	97014	1973484	1.16
Co-krigung	(AE)	371	325	283	1456	0.81
	(SE)	240767	385760	80241	2119192	1.25

Fig. 17.6. Difference of variance between point kriging and point co-kriging. (Leenaers et al. 1989) (Reprinted from Geostatistics, Vol. 1, pp. 371–382, with kind permission from Kluwer Academic Publishers. Copyright 1989. All rights reserved)

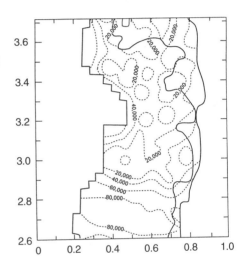

17.4
Limitations of Linear Model of Co-regionalization

The linear model of co-regionalization is a method for modeling the auto- and cross-variograms of two or more variables so that the variance of any possible linear combination of these variables is always positive (Isaaks and Srivastava 1989). Each variable is characterized by its own sample autovariogram and each pair of variables by their own sample cross-variogram. The model for each of these sample variograms may be a composite of one or more basic models, but the same basic model must appear in each auto- and cross-variogram, i.e., each auto- and cross-variogram model must be constructed using the same basic variogram models (Isaaks and Srivastava 1989). Some restriction on the linear model equations concerning positive definite matrices poses restrictions on the modeling by co-regionalization and this makes modeling more difficult. Often one of the auto- or cross-variogram models may not fit its sample variogram very well, leading to a choice of individual models that relates to a good overall fit. Also, a basic model that appears in any auto variogram model does not necessarily have to be included in the cross-variogram model, and secondly, any basic model that is included in the cross-variogram model must be necessarily included in all the auto-variogram models. The equations and matrices for modeling a dataset is found in Isaaks and Srivastava (1989) and the reader is referred there for further insight.

17.5
Analysis of Directional Data

The treatment of directional data is very important in the context of geological analyses and spatial analysis but will be addressed in a brief manner. The reader is referred to Mardia et al. (1979) for a fuller treatment of the theory and examples of application. Mardia et al. (1979) completed a test of preferred and mean direction using principal components analysis. The analysis of directional data is often done through a likelihood ratio test (LRT). These tests include a test for mean direction (likelihood ratio test, Wilks' lambda) and testing the uniformity of axial data (LRT Rayleigh test; see Mardia et al. 1979).

17.6
Summary

The theory of regionalized variables has been successfully used in studying hydrologic parameters and in mineral assays. The techniques are applicable to many multivariate problems in science and, especially, geohydrology.

17.7
Supplemental Reading

Marsily G de (1986) Quantitative hydrogeology. Academic Press, San Diego
Mardia KV, Kent JT, Bibby JM (1979) Multivariate analysis. Academic Press, New York
Tabachnick BG,Fidell LS (1989) Using multivariate statistics, 2nd edn. Harper and Row, New York

Multivariate Data Preparation, Plotting, and Conclusions

CHAPTER 18

Multivariate Data Preparation and Plotting

18.1
Introduction

Many types of data plots exist for exploring geohydrologic and other scientific data. Several types will be shown and described in the next sections.

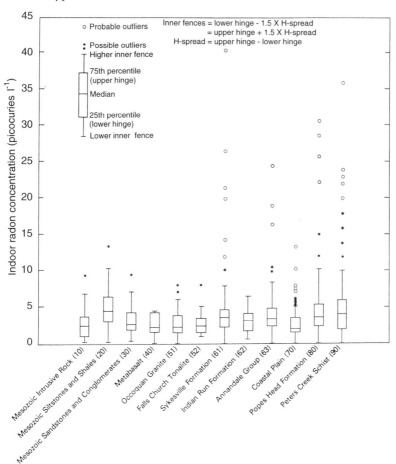

Fig. 18.1. Box plots showing winter indoor radon by geologic unit. (Brown et al. 1992) (Reprinted from Environ. Geol. Water Sci., with kind permission of Springer-Verlag. Copyright 1992. All rights reserved)

18.2
Box Plots – Radon Concentrations

Box plots are useful and concise graphical displays for summarizing the distribution of a data set. Box plots display the main feature of a data group (or level) and are often used instead of histograms. Many statistical analysis programs exist to make box plots and these texts should be consulted (Minitab, Inc. 1986 or other preferred statistical package).

Box plots may be used to distinguish differences in measured values of multiple populations. Brown et al. (1992) used box plots to summarize radon data from a suite of rocks in Virginia. These plots showed differences in radon measured in homes that are located on different rocks with different geochemical and rock properties (Figs. 18.1 and 18.2).

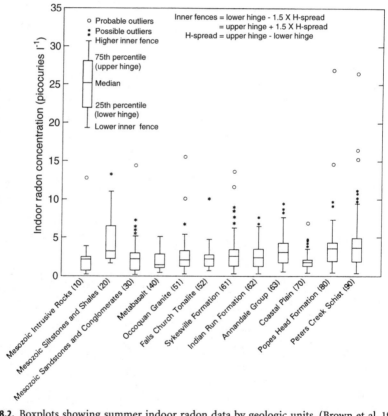

Fig. 18.2. Boxplots showing summer indoor radon data by geologic units. (Brown et al. 1992) (Reprinted from Environ. Geol. Water Sci., with kind permission of Springer-Verlag. Copyright 1992. All rights reserved)

18.3
Scatter Plots – Carbonate Rock Data

Scatter plots are often used to ascertain linearity among pairs of variables and whether transformation may be needed if nonlinearity is found in the data relationships. Scatter plots are valuable plots for inspecting all types of data for population characteristics such as outliers, heteroscedasticity, or other patterns.

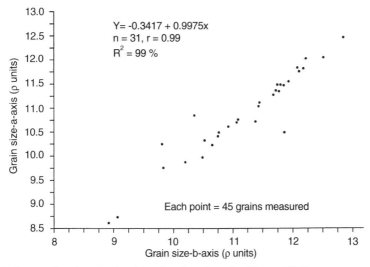

Fig. 18.3. Scatter plot of grain-size data of carbonate rocks. (Brown 1977)

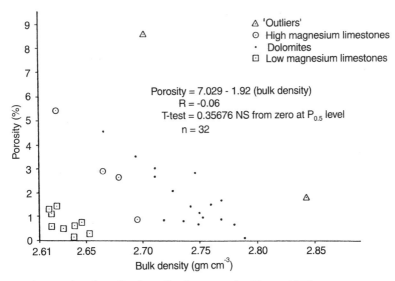

Fig. 18.4. Scatter plot of aquifer data of carbonate rocks. (Brown 1977)

Figures 18.3 and 18.4 show the data plotted for variables in a study on carbonate rocks. The plot indicated that at least two populations existed in the data. This was later confirmed through a cluster analysis. Grain size of long axis (a) and short axis (b) of carbonate grains were shown to exhibit a high degree of linear association.

18.4
Blob Plots – Geographic Data

A blob plot is a graphical plot of more than one variable. The plot is simple and has the axes labelled and plotted accordingly, and can be used with other multivariate methods. The size of the blob, and type of symbol is arbitrary but often an octogon or square is used as the plotting symbol. Figures 18.5 and 18.6 show a blob plot for geographic data from PCA analysis and air pollution data, respectively (from Johnson 1987).

18.5
Multidimensional Data Plot – Toxicological Data

One of the most useful graphs for exploring data is a multidimensional plot of the multivariate data. In a study by Dixon and Sprague (1981)), dosage – response data were plotted on a multidimensional graph. Howarth and Sprague (1978) plotted toxicity data on a multidimensional graph to illustrate effects of concentrations of total dissolved copper on rainbow trout (Fig. 18.7). The lethal-

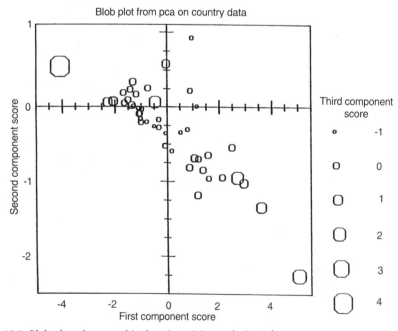

Fig. 18.5. Blob plot of geographic data from PCA analysis. (Johnson 1987)

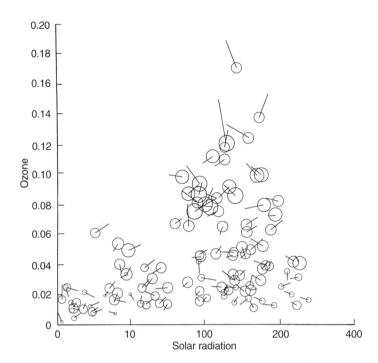

Fig. 18.6. Blob and metroglyph plot of air pollution data. (Johnson 1987)

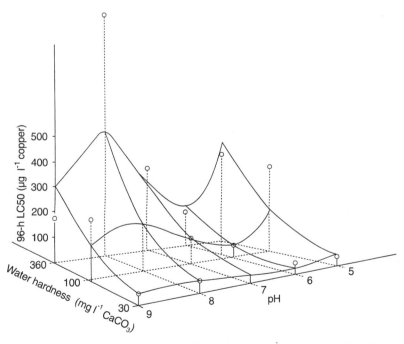

Fig. 18.7. Multidimensional data plot of chemical data for toxicity tests. (Howarth and Sprague 1978) (Reprinted from Water Research, Vol. 12, p. 455–462, with kind permission of Elsevier Science Publishers, Oxford OX5 IGB England. Copyright 1978. All rights reserved)

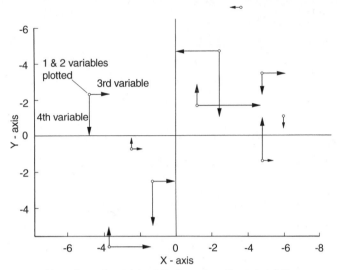

Four-dimensional data plotted on a two-dimensional diagram

Fig. 18.8. Plot of multidimensional data

ity of copper to the trout was clearly illustrated by this very simple plot. Another multidimensional plot is shown in Fig. 18.8.

18.6.
Metroglyphs

Metroglyphs are two-dimensional graphs that use arrows of different sizes to depict the multidimensional nature of data (Fig. 18.9). Many dimensions can easily be shown using this simple plot.

18.7
Chernoff Faces

Chernoff faces are graphs that depict the characteristics of data by using facial parts of different sizes to represent the multidimensional nature (Fig. 18.10). Enlarged facial parts are used to illustrate increased size of variable values and smaller facial parts for smaller values of variables. These simple plots thus depict characteristics of multivariate data using a face depiction in which parts of the face take on different sizes based on measured variables being analyzed.

18.8
Andrews Plots

Andrews plotting is a technique that uses Fourier analysis to transform the results of multivariate data and for representing a set of multivariate data by a

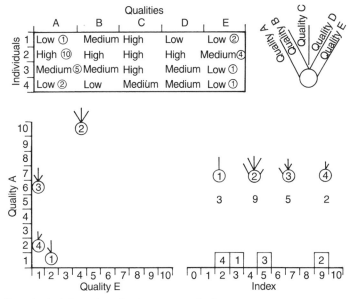

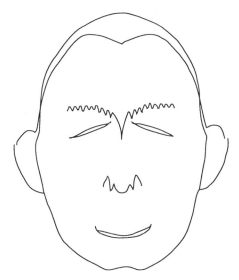

Fig. 18.9. Hypothetical data plotted using a metroglyph

Fig. 18.10. Chernoff face of hypothetical data

Variable cartoon face features
1. Upper hair
2. Chin curve
3. Lower hair
4. Eye size
5. Mouth size
6. Eye space
7. Eye slant
8. Mouth curve
9. Face size
   (eyes to moth)

wave form pattern. The wave form pattern can show which variables and sample groups are more similar or which sample groups exhibit the same trends in the data (Fig. 18.11).

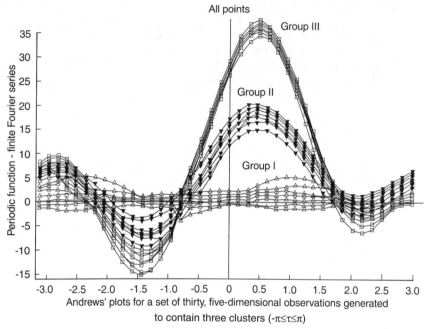

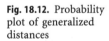

Fig. 18.11. Andrews plot of hypothetical data

Fig. 18.12. Probability plot of generalized distances

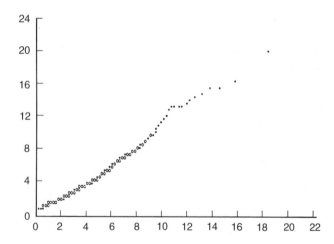

18.9
Probability Plots

Probabilty plots are often used to ascertain if data fit a normal distribution. When data are plotted, the data will approximate a straight line when the data are taken as fitting a normal distribution (Fig. 18.12).

18.10
Histograms – Carbonate Rock Data

Histograms are very important graphs for ascertaining asymmetry in the measured data. From these graphs, it is possible to make simplistic judgements on aspects of normality, and further tests can be executed using chi-square tests for normality. Histograms for carbonate rock data were completed for measurements from a suite of carbonate rock aquifers in central Pennsylvania (Fig. 18.13 and 18.14).

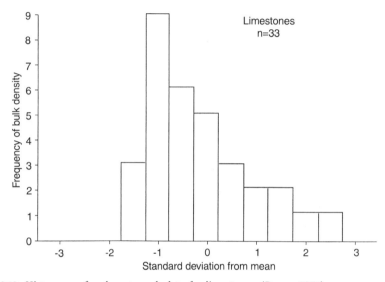

Fig. 18.13. Histogram of carbonate rock data for limestones. (Brown 1977)

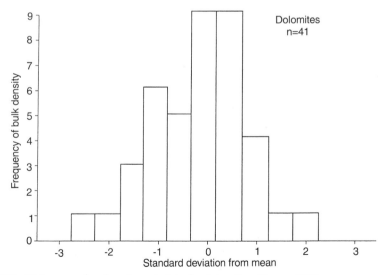

Fig. 18.14. Histogram of carbonate rock data for dolomites. (Brown 1977)

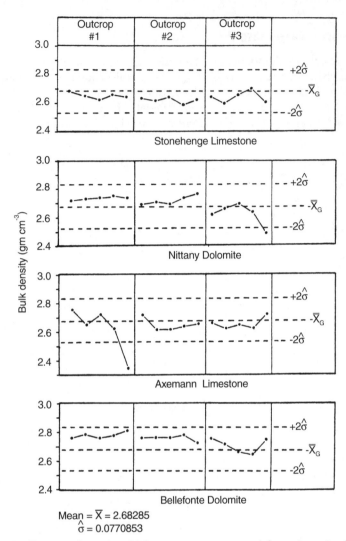

Fig. 18.15. Quality control graphs which compare outcrops and formations simultaneously. (Brown 1977)

18.11
Quality Control Graphs – Carbonate Rock Data

Quality control graphs are used in industrial technology studies and other scientific studies to ascertain when experiments may be out of control, i.e., when parts may need replacing or when some process has changed. In geological studies, it is possible to plot rock data and see how populations compare based on the variability in individual samples, outcrops, or formations. Brown (1977) completed quality control graphs to compare measurements on carbonate rock aquifers in central Pennsylvania (Figs. 18.15 and 18.16).Variations in the outcrops and formations are clearly shown in these graphs.

18.12
Star Diagrams

Star diagrams, similiar to rose diagrams in geological studies, show size of variables on axes that have been labelled according to scale of measurement. An example is a rose diagram in geological analyses of strike and dips. Other examples are given in Johnson and Wichern (1988).

18.13
Simultaneous Confidence Intervals and Bonferroni Intervals

Bonferroni intervals are alternative comparison methods that use T-square procedures. They are used to compare mean vectors. The reader is referred to Johnson and Wichern (1988) for a better treatment of this topic and examples of confidence ellipses.

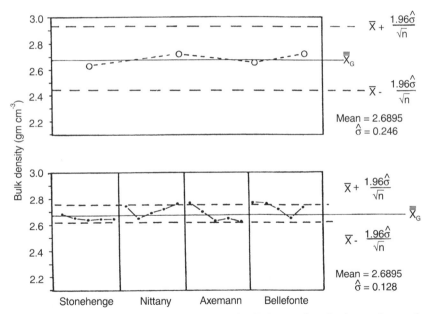

Fig. 18.16. Quality control graph for comparison of bulk density data for four carbonate formations. (Brown 1977)

18.14
Summary

Multivariate data plotting is a vital part of any multivariate analytical study. This chapter has defined most of the available methods that are widely used for graphing multivariate data.

18.15
Supplemental Reading

Johnson RA, Wichern DW (1988) Applied multivariate statistical analysis. Prentice Hall, Englewood Cliffs

CHAPTER 19

Summary and Generalizations of Multivariate Quantitative Procedures

19.1
Introduction

Research problems lend themselves to analysis by a number of multivariate statistical applications. This text has attempted to address some of the applications and underlying basis for choosing a procedure or technique. Scientific parsimony is the essence of scientific reasoning and should always be adhered to. The explained variance in a dataset is a significant aspect of any structural analysis whether it be the explained variance of a row, explained variance of a group of measurements, or the explained variance of a factor or column. In any situation, if we can reduce the number of "explainer" variables or reduce the number of measurements to attain a true value or estimate, we have fulfilled the task of multivariate statistics. The examples chosen for review represent a broad spectrum of methodologies and techniques, by which scientific data can be parsimoniously analysed using model building and model fitting strategies.

Reasoning in the sciences is accomplished in many ways and hence uniqueness as well as nonuniqueness is of essence and has value as research is completed. The style of the scientist is an academic choice but the consequence of the choice should be analysed also. It is hoped that the salient basis of all the choices in the sciences is parsimony.

19.2
Multivariate Methods and Generalizations

Multivariate data results from measuring several different variables on each experimental unit or sampled object. A variable is a characteristic of a population that can have different values. Almost all data are multivariate in scope and such data frequently occur in all branches of science. The multivariate methods are all based on the techniques of matrix algebra and this supporting mathematics prevents widespread miscommunication in terms of methods used.

19.3
Methods Covered

Most of the discussions in this text cover such general multivariate topics as factor analysis, principal components, correlation, cluster analysis, regression, and less commonly used techniques such as discriminant analysis, multivariate analysis of variance (MANOVA), canonical correlation, canonical variate analysis, multivariate analysis of covariance (MANCOVA), multivariate data plotting, and profile analysis. This list by no means exhausts the possible procedures that can be used in analyses.

19.4
Normality Tests

Graphical displays of data such as histograms and probability plots can be very important in recognizing how data are distributed and if gross errors in the data exist. It is very important to determine when data is nonnormal. New methodologies for normality tests now include using probability plots or goodness of fit tests such as chi-squared tests.

19.5
Multivariate Data Preparation and Plotting

The data for multivariate analysis should always be examined for normality, homogeneity of variances, and multicollinearity. These examinations should be directed at determination of the suitability of the data for analysis, and, if transformations are necessary, what form the data should take. In the context of data analysis, it is important to both prepare the data and examine the data. In some instances, new techniques in simple statistical analysis may be used to explore data for hidden trends. These exploratory data techniques include boxplots, scattergrams, histograms, stem-and-leaf plots, and probability plots, to name a few. The test for normality may be a normal probability plot on variables, tests of skewness and kurtosis, chi-square goodness of fit tests, and/or histograms.

The homogeneity of variances tests are also evaluated often by using Bartlett's test of homogeneity of variances. Transformations are assumed appropriate for the data by using the ladder of powers method, but often the transformation used is simply the logarithm. Transformations are used to make data more linear, more symmetric, or to achieve constant variance.

19.6
Correlation Analysis

The correlation coefficient measures the strength of the linear relationship between variables $x(i)$ and $x(j)$. Coefficients close to zero indicate that no linear relation exists between the two variables, but they do not necessarily imply that no relation exists, unless the two variables have a joint bivariate normal distribution. Many software packages give tests of each correlation. When N is very large, virtually all sample correlation coefficients will be significantly different from zero, so the test is not very interesting. The important question to ask is: are they important? The determination should be made as to which correlations are of practical importance. As a rough guide correlation coefficients whose absolute value is larger than 0.7 may be considered large and important to the analysis.

A logical approach to analysis of multivariate data is begun with the reduction of the [n × p] data matrix to a [p × p] matrix of correlation coefficients. A basic feature of this correlation is that the values of the variables have been transformed so that each mean is zero (mean X = 0) and the variance is equal to one (variance = 1). This permits a quick comparison of the different measured variables. If significant correlations in the data are not found during this preanalysis phase, it is not necessary to proceed to principal components analysis or factor analysis and a new suite of variables must be determined and analyzed. Principal components analysis and factor analysis assume that intercorrelations do exist in the data. Use of correlation coefficients requires testing of the significance of the statistical estimators before drawing inference, and this approach usually pertains to use of special statistical tables based on degrees of freedom. Correlation coefficients should not be calculated for sample size N < 12, because results will be unreliable and spurious at best.

When pairs of variables are highly correlated, one should partition the response variables into groups so that variables within a group have high correlations with one another and variables in different groups have low correlations with one another. Such a partitioning may reveal important aspects of the data which will often be useful in deciding how to interpret it.

19.7
Coefficient of Variation

If the absolute dispersion is the standard deviation, sd, and the average is the mean, the relative dispersion is called the coefficient of variation (CV) or coefficient of dispersion and is the ratio of sd to the mean. The relationship between mean and dispersion is very important in the geosciences and is expressed by the coefficient of variation, CV % = 100 sd/mean. The coefficient of variation is attractive as a statistical tool because it apparently permits the comparison of variates free from scale effects; i.e., it is dimensionless. However, it has appropriate meaning only if the data achieve ratio scale. The coefficient of variation can be plotted as a graph. A CV exceeding, say, about 30 % may show that an experiment is out of control, or that sample groups are very different in their measured properties.

19.8
Factor Analysis

Factor analysis (FA) is more concerned with explaining the covariance structure of the variables, whereas PCA is more concerned with explaining the variability in the variables. In studying correlations between variables, one finds a systematic decrease in correlations in each row, more or less to the same extent. The salient point is that this could be accounted for if the score on the ith variable were composed of two parts, a random component common to all variables and random part specific to each variable: i.e., $X_i = \lambda_i Y + E_i$, where Y = a random component common to all variables, and E_i = random component specific to each variable. This is the basic model for factor analysis.

Since the factor loading matrix is not unique, the custom has been to rotate the factors obtained in order to select the loadings in such a way as to make more sense. Only the significant factors should be rotated, and the others factors discarded. Most rotation procedures try to make as many loadings as possible near zero and to maximize as many of the others as is possible. Also, since the factors are independent, it would be nice if a particular variable is not heavily weighted on more than one factor. There are many methods of rotation, that tend to keep the factors independent, i.e., uncorrelated when one begins with an independent set of factors.

Factor analysis produces what are called factors and principal components analysis, which is a similar method, produces components. The procedures are similar except in preparation of the observed correlation matrix for extraction. The difference between FA and PCA is in the variance that is analyzed. In PCA, all the variance in the observed variables is analyzed, but in FA only the shared variance (communalities) is analyzed and attempts are made to estimate and eliminate variance due to error and variance that is unique to each variable. Mathematically, the difference between PCA and FA involves the contents of the positive diagonal in the correlation matrix (the diagonal that contains the correlation between a variable and itself). In either PCA or FA, the variance that is analyzed is the sum of the diagonal elements. When the covariance matrix is not known, factor analysis is applied to the sample covariance matrix or the sample correlation matrix.

In the principal-factor method, the user is required to give suitable estimates of the communalities, or more equivalently, estimates of the specific variances. The covariance matrix from the multiple (m)-factor model is the matrix of variances and covariances 'reproduced' and is not to be confused with the sample covariance matrix. The principal-factor method with iteration uses the maximum likelihood method, gives results that do not vary whether one uses the correlation or covariance matrix, and hence is scale invariant. Many other methods have evolved over the years. Factor analysis can be started using the results from a PCA to determine the number of factors that are necessary. Sometimes a cluster analysis is performed on the variables and is forced to produce as many clusters as the m-factors required.

19.9
Canonical Correlation

Canonical correlation is a generalization of multiple correlation. It is relatively a problem of simultaneously reducing two symmetric matrices to diagonal form. Variables are often found to belong to different groups that are generalized to relate to different processes or factors. Canonical correlation analyses are used to identify and quantify the associations between two sets of variables in a data set.

Canonical correlation analyses have the objective of determination of the correlation between a linear combination of the variables in one set and a linear combination of the variables in another set. The first pair of linear combinations have the largest correlation. The second pair of linear combinations are determined and have the second largest correlation of the remaining variable sets. This process continues until all pairs of remaining variables are analyzed. The pairs of linear combinations are called canonical variables, and their correlations are called canonical correlations. The number of canonical correlations is equal to min $(q, p-q)$. The number of nonzero canonical correlations is equal to the rank of the covariance matrix. A relation between canonical correlations and multivariate regression exists, if one thinks of the elements of X_1 as dependent variables and those in X_2 as independent variables, and X is regressed on X_2.

19.10
Multiple Linear Regression

Multiple-regression analysis is used to derive an equation that can be used to predict values of the dependent variable from several independent variables. The problem narrows down to finding the best function of the form represented by the regression equation to predict the mean value of y from the x's, and is often done by least squares estimation which minimizes the residual sum of squares that is done through the solution of the normal equations. This procedure leads to an analysis of variance test of significance for the relationship between y and the x's. Any observed variable can be considered to be a function of any other variables measured on the same samples.

19.11
Nonlinear Regression

For many problems the linear model may not be appropriate, at least as a first approximation to the true underlying model. For other problems a transformation to linearity might suffice for the parameters. There are many situations, however, in which a linear model is not appropriate, for example, if the underlying model is a sum of exponential and/or trigonometric functions. In this case, a transformation to a linear model is not easily done and an appropriate approximation cannot be made. The equation not convertible to linear form is said to be intrinsically nonlinear. A model of this nature is termed a nonlinear regression model, and is solved iteratively.

19.12
Principal Components Analysis

The main objective of a principal components analysis (PCA) is to replace the original variables by a smaller number of underlying variables that explain all of the variability in the data. It is hoped that the first few principal components account for most of the variability in the data so the effective dimensionality of the data is reduced.

The objective of principal components analysis is also to determine the relations existing between measured properties that were originally considered to be independent sources of information. Principal components are the eigenvectors of the variance-covariance matrix, developed from the original data matrix. Through evaluation of the principal components, one seeks to determine the minimum number of variables that contain the maximum amount of information and to determine which variables are strongly interrelated. The physical significance of the interrelations of the components of the data are sought to provide a simple interpretation of processes causing variation in variables.

19.13
Correspondence Analysis

Correspondence analysis is a way of interpreting contingency tables that may be defined through principal components analysis. In correspondence analysis, a factor is represented by the eigenvector of the normalized covariance or correlation matrix. It can be further viewed as a simultaneous linear regression scheme with dual scaling, which allows the interpretation of both sample sites and variables in the same factor space.

19.14
Discriminant Analysis

Discriminant analysis techniques are used to classify individuals into one of two or more alternative groups (or populations) on the basis of a set of measurements. The populations are known to be distinct, and each individual belongs to one of them. These techniques can also be used to identify which variables contribute to making the classification. Thus, as in regression analysis, we have two uses, prediction and description.

Sometimes discrete variables may be incorporated using dummy variables as is done in dummy variable regression. The Mahalanobis distance is a distance measure that is adjusted for the variances and covariances of the responses and is used in discriminant analysis.

19.15
Multivariate Analysis of Variance (MANOVA)

The consideration of procedures based on normal distribution theory have been extended to the analysis of data arising from designed experiments. The calculation of a multivariate analysis of variance (MANOVA) is essentially similar to the calculation of a univariate analysis of variance (AOV). However, the testing of hypotheses and the interpretation of results are more complicated.

19.16
Multiple-Factor Analysis of Variance (Factorial Design)

Multi-factor analysis of variance involves the study of more than one factor. What this means is that the design incorporates more than one independent variable, and each of these variables or factors can have two or more levels of its own. Usually we call such approaches 'factorial designs' and describe them in terms of the number of factors and the number of levels each has. For example a 2×2 design incorporates two factors each having two levels. A $4 \times 2 \times 2$ design has three factors, the first having four levels, the second factor having two levels, and the third factor having two levels. In a given design, factors and their levels define different subgroups in the experiment. Multiple-factor analysis of variance or factorial AOV provides methods for testing if different subgroups, or various combinations of subgroups, represent different populations in terms of what is being measured as the dependent variable. One of the useful features of a factorial AOV is that it allows us to test a number of different hypotheses in a single study.

19.17
Cluster Analysis

Cluster analysis can be run on samples or variables and results are easily explained. Cluster analysis is a separate and useful technique for grouping individuals or objects into unknown groups. It differs from other methods of classification, such as discriminant analysis in that in cluster analysis the number and characteristics of the groups are to be derived from the data and are not usually known prior to the analysis, thus it is not a true classification technique.

Variable cluster analysis is sometimes used as an alternative to factor analysis, and begins with all variables in one group or cluster. The first two principal components are found, then an orthoblique rotation is done and variables are assigned to the rotated component with the higher squared correlation. This process continues and variables are iteratively reassigned to groups to maximize the variance accounted for by the group components.

19.18
Multivariate Analysis of Covariance

Multivariate analysis of covariance (MANCOVA) is the multivariate extension of analysis of covariance that is discussed earlier in this text (see Chap. 8). MANCOVA addresses the hypothesis of whether there are statistically reliable mean differences among groups after adjusting the newly created dependent variable for differences on one or more covariates. A good summary that discusses the major concepts of the analysis of covariance has been given earlier to introduce the subject of multivariate analysis of covariance and its possible use in geohydrology.

19.19
Analysis of Covariance

The analysis of covariance (ANCOVA) is commonly done as an analysis of variance and regression analysis combined. For the analysis of covariance to be an improvement over the analysis of variance, the common slope (B) must not = 0, as such we test the slope with a F-test before proceeding to ANCOVA. The analysis of covariance is often used as a procedure to adjust the analysis for variables that could not be controlled by the experimenter. As used in this text, the analysis of covariance can be used as a procedure for comparing several regression lines or surfaces, one for each treatment or treatment combination, where there is possibly a different regression surface for each treatment or treatment combination. There are many research situations in which one covariate is measured on each experimental unit, giving rise to the model equation for each treatment.

19.20
Special Tests on Covariance Matrices

The tests on covariance matrices of most utility cover:

1. tests on the structure of a matrix;
2. tests about independence or sphericity;
3. tests as to whether submatrices equal zero;
4. tests as to whether covariance matrices are equal.

A likelihood ratio test is often employed to complete these tests.

19.21
Multivariate Multiple Regression

Multivariate multiple regression is used when we want to regress a set of dependent variables $X(1)$ on a set of independent variables $X(2)$. The treatment of this topic is not widespread in the literature and the reader is referred to Johnson and Wichern (1988) or other texts for a fuller treatment and background. The canonical correlation method has many similiar goals.

19.22
Logistic Regression

Logistic regression modeling is the preferred choice for data analysis over discriminant analysis when multivariate normality of the data is not apparent and an individual is to be classified into one of two populations. Thus it is an alternative to discriminant analysis; but if the data is multivariate normal, a smaller set of data is required by discriminant analysis to achieve the same precision as the logistic regression model. This technique has been widely used in the medical field and industry to predict success or failure based on several measured variables.

Logit analysis with a dichotomous dependent variable can be performed through logistic regression, and logistic regression is the more general procedure because it allows continuous as well as categorical independent variables. The term 'logit' refers to the interpretation of the parameters as the log of odds ratio.

19.23
Multivariate Probit Analysis

Probit analysis is defined as a procedure to study the dosage-response relation in a population of organisms, where randomly chosen population members are exposed to various levels of applied stimulus and quantal response is assessed as either dead or alive. In some instances more than one organism or physiological system is affected by the stimulus leading to a test of so-called main effects and side effects.

19.24
Multivariate Time Series Analyses

The analysis of multivariate time series modelling was not treated in detail as only a general introduction was given in previous discussions. The methods that are commonly used to describe multivariate time series include: (1) dynamic regression methods that link time series and causal affects; (2) Box-Jenkins autoregressive integrated moving average (ARIMA) methods that use rational polynomials for parsimonious representation of the autocorrelation function of a time series; and (3) state space methods that are statistically the same or similiar to Box-Jenkins (ARIMA) but are truly multivariate and use canonical correlation theory to determine the mode and identify the model parsimoniously. The above methods usually employ some type of information criterion such as: (1) Aikaike AIC, which balances goodness of fit to the historical data and model complexity and is used across methodologies, or (2) Schwarz BIC that is similiar to AIC but penalizes complexity more severely in model selection. For other discussions of concepts in time series analysis, many textbooks exist and the reader is referred to these.

19.25
Multidimensional Scaling

This is a technique for reducing dimensionality and is based on the distance between points. The result is a reduction in the dimensions used to represent the data with as little distortion as possible. It is similiar in its objective to principal components analysis discussed earlier. Studies concerning the analysis of distances between places on a map have received much attention in multidimensional scaling applications.

19.26
Multivariate Spatial Statistics

The underlying mathematical basis for looking at spatial statistics in the geosciences derives from the theory of 'regionalized variables' i.e., those variables that have some degree of spatial correlation. The key point in this theory is that a geological process active in the formation of a mineral deposit or any other chemical or fluid body, is interpreted as a random process. Examples of regionalized variables are grade of ore, thickness of a formation, or elevation of the surface of the earth, to list a few. Because the formation of the body is a random process, the grade or concentration at any point in a deposit or fluid body, is considered as a particular outcome of this random process. This probabilistic interpretation of a natural process as a random process is necessary to solve the practical problems encountered in estimating geologic or hydrologic variables. The method assumes that the adjoining samples are correlated to one another, which directly opposes ideas of classical statistics. It assumes that the particular relationship expressing the extent of the above correlation can be analytically and statistically captured in a function known as the variogram function or variogram. The function $Z(x)$ displays a random aspect consisting of highly irregular and unpredictable variations, and a structured aspect reflecting the structural characteristics of the regionalized phenomena.

A logical extension to the situation where two or more variables are spatially interdependent and the one is undersampled is a method called co-kriging. Co-kriging is useful for interpolating a property that is more difficult and costly to measure. Co-kriging makes use of the large spatial correlation between the property of interest and one of the less difficult to measure properties. The term "co-regionalization" analysis has been used to describe the analysis of co-regionalization matrices obtained from multivariate spatial data using a nested variogram model.

19.27
Nonparametric Multivariate Analysis

The concept of nonparametric multivariate analysis is based on studies by Mantel and Valand (1970). On each of n individuals, or samples, $p + q$ variables are measured. A nonnegative distance or closeness measure between any two individuals on any one variable can be based on ranks or tied ranks for orderable variables (continuous, discrete, or categorical); for nonorderable categorical variables the distance measure reflects whether the two individuals belong to the same category. A test statistic is used to judge closeness.

19.28
Summary

This chapter has presented a short summary of all the different procedures discussed in this text. It can be used as a reference for utilizing various techniques in future investigations.

All geohydrologic data should be graphed in some fashion to look at trends and data distribution attributes. This can best be carried out with one or several of the methods summarized in this chapter.

19.29
Supplemental Reading

Johnson RA, Wichern DW (1988) Applied multivariate statistical analysis. 2nd edn. Prentice Hall, Englewood Cliffs
Mardia KV, Kent JT, Bibby JM (1979) Multivariate analysis. Academic Press, New York
Tabachnick BG, Fidell LS (1989) Using multivariate statistics. Harper and Row, San Francisco

Introduction to Numerical Analysis

A.1
General Concepts

A short review is provided for those readers that do not have access to mathematical references giving the simple mathematics needed for this text. Many texts are however available that treat this area of mathematics.

The ultimate aim of numerical methods is to provide a convenient way of obtaining useful solutions to mathematical problems and for extracting useful information from available solutions which are not expressed in tractable form. Generally it is not the aim of numerical methods to obtain exactness. Their goal is primarily to provide approximations which differ from the exact one by less than a specified tolerance, or, statistically speaking, by a value which has less than a specified probability of exceeding it.

A.2
Solution of Simultaneous Linear Algebraic Equations

A.2.1
Sets of Linear Equations

Simultaneous equations occur widely in many fields of engineering and science. In many cases they are used to describe directly the physical problem, but often the solution of a system of linear equations is some larger problem. One way of considering simultaneous equations is if one considers a set of n equations, describing relationships among n unknowns, $x_1, x_2, \ldots x_n$, in the form:

$$a_{11}x_1 + a_{12}x_2 + \ldots + a_{1n}x_n = b_1$$
$$a_{21}x_1 + a_{22}x_2 + \ldots + a_{2n}x_n = b_2 \qquad (A.2.1.1)$$
$$a_{n1}x_1 + a_{n2}x_2 + \ldots + a_{nn}x_n = b_n,$$

where each term in each equation contains only one unknown x_i. These are to the first power only and hence the equation is linear. The first subscript describes the row and the second denotes the column to which a_{ij} is assigned. The solution sought represents a set of values for the n unknowns which, when substituted into Eq. (A.2.1.1) above satisfies all of them simultaneously. The geometric interpretation of a linear algebraic equation is a straight line in the case of two unknowns, a plane for three unknowns, and a hyperplane for four or more unknowns. The above equation can be written in the form:

$$\sum_{j=1}^{n} a_{ij} x_j = b \quad (i=1,2,3,\ldots,n) \tag{A.2.1.2}$$

In the case where all the right-hand members, b_i are zeros, the set of simultaneous equations are said to be homogeneous, and as such, is always solvable. In fact, one solution is always the trivial one, i.e., $x_1 = x_2 = x_3 = \ldots = x_n = 0$. In matrix notation, this can be written as: $[AX = B]$, where A represents the square array of the coefficients a_{ij} known as the coefficient matrix written as:

$$[A] = \begin{matrix} a_{11} & a_{12} & a_{1j} & a_{1n} \\ a_{21} & a_{22} & a_{2j} & a_{2n} \\ a_{i1} & a_{i2} & a_{ij} & a_{in} \\ a_{n1} & a_{n2} & a_{nj} & a_{nn} \end{matrix} \tag{A.2.1.3}$$

and $[X]$ = column matrix of x_i variables i.e., x_1 to x_n, and $[B]$ = column matrix of right-hand side constants b_i, (i.e., b_1 to b_n). In this form where we separate the coefficients from the variables, the column of variables are called vectors, and the array of coefficients a m x n matrix. From a given arbitrary set of linear algebraic equations, there are three possibilities:

1. There is a unique solution;
2. There is no solution; in this case a system of equations are said to be singular and ill-conditioned. What is important here is that numerical methods for the solution of linear algebraic equations (such as Gaussian elimination) give immediate information about the singularity of a system of equations.
3. There is an infinite set of solutions; this set is also said to be singular, a condition which must be determined before attempting the actual solution of a set of equations.

In the theoretical sense, a set of linear algebraic equations are either singular or not, but in the case of numerical computation, it can be near-singular, which leads to an unreliable solution. For an ill-conditioned set of equations, the problem usually lies in the fact that the lines representing the the equations are nearly parallel, and it is difficult to find the numerical solution, i.e., the point of intersection with reliable accuracy. In the case of three or more dimensions (system of three or more unknowns) the situation is further complicated by the fact that there are ways that a set of linear algebraic equations can be singular or near-singular without any planes or hyperplanes being parallel or nearly parallel. While the use of determinants has drawbacks, the best methods are: (1) direct or finite methods and (2) indirect or infinite methods that use some form of iteration procedure.

A.2.2
Calculations in Matrix Algebra

Most of the algebraiz manipulations with which the reader is familiar – addition, subtraction, multiplication, and division – have counterparts in matrix algebra.

These calculutions are commonly done on square matrizes, i.e., sums-of-squares and cross-products matrices, variance-covariance matrices, and correlation matrices.

A.2.3
Definitions and Notation

A large part of scientific computations involves matrices in some form, and includes concepts such as matrix inversion, relaxation techniques, eigenvalues and eigenvectors. A summary of some of these concepts is important and will be presented now.

Definition A.2.3:

A matrix is a rectangular array of quantities. A matrix has m rows and n columns and is denoted by: [A] = same as Eq. (A.2.1.3.) If m = n, the matrix is said to be square of order n; if m = 1, the matrix has one row and is called a row-vector. Further, if n = 1, the matrix is called a column-vector. If m = n = 1, the matrix reduces to a scalar quantity a(1). The main or principal diagonal of a square matrix is the string of elements from the upper left corner to the lower right. The other diagonal is called the secondary diagonal.

The determinant of a matrix is not a matrix, but is a scalar quantity computed from the elements of an array. A diagonal matrix is a square matrix with zero elements off the main diagonal. An identity matrix is a diagonal matrix with ones on the main diagonal and may be denoted as [I]. If the order is also known for the identity matrix, it can be also noted as [I(4)]. A zero matrix is a matrix with all its elements zero.

A.2.4
Basic Matrix Operations

A.2.4.1
Comparison of Matrices

Only matrices of the same dimensions (identical number of rows and columns, and corresponding elements a_{ij} and b_{ij} are equal for all i, j) can be compared, and are said to be equal for all i, j.

A.2.4.2
Matrix Addition and Subtraction

For matrix addition and subtraction, matrices must have the same dimensions, and this characteristic allows corresponding elements to be operated upon giving; [A] + [B] = [C] or [A] – [B] = [C]. Coordinate translation of axes in three dimensional Euclidean space is accomplished by addition and subtraction of matrices.

A.2.4.3
Matrix Multiplication

In matrix multiplication, the order in which the two matrices are written is important because [A] × [B] may not equal [B] × [A], and the matrices may be said to be pre-multiplied or post-multiplied. The number of columns in the lead or first matrix must equal the number of rows in the second or lag matrix. Elements of the product matrix are said to be scalar products of the ith row and jth column of two matrices operated upon.

The rotation of a point about some coordinate axis in Euclidean space is done through matrix multiplication. Examples may be found in other texts (Davis 1986). Division is replaced by either premultiplication or postmultiplication by the inverse matrix.

A.2.4.4
Transposition

The tranpose of a matrix ($[A]^T$) is the result of interchanging rows and columns in a matrix.

A.2.4.5
Partitioning

A matrix may be partitioned horizontally or vertically into a number of sub-matrices. Each submatrix then becomes an element of the partitioned (original) matrix and is treated according to other elements in a matrix in terms of basic operations. In particular, partitioning of a matrix simplifies multiplication when the matrix contains special submatrices such as the identity matrix or zero-matrix. In this instance, multiplication can only be carried out if the lead matrix (first) is partitioned vertically in exactly the same way as the lag-matrix (second) is partitioned horizontally, or vice versa. This is called "conformable partitioning". In the case of a matrix with many zeros and ones, the procedure is to partition into as many identity and zero submatrices as possible. Again, examples may be found in Davis (1986) or Pall (1971).

A.2.5
Matrix Inversion

A.2.5.1
Inverse

There is an analogy between a reciprocal of a number and the inverse of a matrix. In matrix algebra, division (which is not commutative, hence not defined) is achieved by multiplying by the inverse of the divisor matrix. The inverse of a square matrix, [A], if it exists, is another matrix denoted $[A]^{-1}$, such that $[A]([A]^{-1}) = ([A]^{-1})[A] = I$, the identity matrix. Only a square matrix whose rows and columns are independent of each other has an inverse, and is called a non-singular matrix.

A.2.5.2
Computation of Inverse

The computation of the inverse, called inverting, is fairly straightforward. Given a square matrix [A] of order n with elements a_{ij}, the corresponding identity matrix of the same order is I(n). Elementary row transformations are done to the original matrix [A] in order to reduce to the identity matrix I(n). Whenever a transformation is applied to [A], it is also applied to I(n). When [A] has been reduced to the I(n) matrix, the original I(n) matrix will have been built up to $[A]^{-1}$ matrix, which is the inverse of [A]. The elementary transformations are: (1) multiply or divide any row by a constant; (2) add or subtract any row to or from any other row; and (3) interchange any rows as necessary.

A.2.6
Eigenvalues and Eigenvectors of a Matrix

A matrix can be considered as a transformation of a point or of a vector (directed line segment) joining the origin with that point in Euclidean space. Some vectors will not be changed in direction by a given matrix, but only in length. These are called eigenvectors of that matrix and are of special interest in scientific applications.

Eigenvalues and eigenvectors are often desired as part of the solution of partial differential equations. For example, they may specify vibration nodes and resonance frequencies in vibration analysis or other problems. In such cases, the matrix involved is usually real and symmetric. Sometimes only one (such as the largest) eigenvalue is required; sometimes all are required. This will determine the choice of computational methods, but highly effective algorithms such as the Jacobi method will generate all the eigenvalues and eigenvectors for a symmetric matrix. Given an m-element column vector [X], the square vector [A] of order m, and some scalar constant λ, then [X] is said to be an eigenvector of [A], and λ is its associated eigenvalue. Accordingly,

$$\text{if} \quad |[A] - \lambda I| = 0 ; \tag{A.2.6.1}$$

$$\text{then} \quad [A][X] = \lambda[X]. \tag{A.2.6.2}$$

The first part of Eq. (A.2.6.1) which involves a polynomial of degree m in λ, is called the characteristic equation of the matrix [A]. There are m distinct eigenvalues for a matrix of order m. The eigenvalues for a matrix with real elements need not be real but can also occur as pairs of complex conjugate numbers. The eigenvalues for a real symmetric matrix are real, and in general $[A]^T$ (transpose of [A]) has the same eigenvalues (also called characteristic roots, or proper values) as [A]. A normalized eigenvector is a vector that has a length of 1. Eigenvectors are not unique, and it can be shown that if $[A][X] = \lambda[X]$, then $[A](k[X] = \lambda(k[X])$ for any constant k, and there is an infinite number of eigenvectors associated with the eigenvalue λ.

A.2.7
Matrix Algebra and Solution of Simultaneous Linear Equations

The solution of simultaneous linear algebraic equations by means of classical methods using the determinant of the coefficient matrix, such as Cramer's rule, are highly inefficient in actual numerical cases, because of the excessive computational work involved in the evaluation of (n + 1) determinants of order n, when n is very large. Therefore numerous other methods have been divised for computers. One of these is the inversion of the coefficient matrix [A] to obtain the solution in the form of a vector [X]. Remember that matrices with no inverse are singular, and if the coefficient matrix [A] is singular, the corresponding set of simultaneous linear equations is also singular, i.e., it has either no solution or an infinite number of solutions.

A.2.8
Direct Methods: Gaussian Elimination

One of the most efficient ways of solving simultaneous linear equations by direct methods is that of Gaussian elimination. It consists of reducing the set of n equations to triangular form by successively eliminating one unknown at a time. Thus, first the unknown x(1) is eliminated from the last (n –1) equations; then x(2) is eliminated from (n – 2) of the (n – 1) equations not containing x(1); and the process is continued until, in the last stage of elimination, the last (nth) equation of the set assumes the form x(n) = constant. The equation is now solved for x(n) and the result is substituted back into the preceding equation to solve for x(n – 1) until the back substitutions result in solving for all variables x(i). This means that Gaussian elimination in general comprises two distinct phases: the forward elimination and back substitution. The formula for the kth stage of elimination is the heart of the algorithm itself.

A.2.9
Indirect Methods: Gauss-Seidel Iteration

Gauss-Seidel iteration is a fairly straightforward iterative procedure that is good for large systems and is discussed in many other texts on numerical methods (Wang and Anderson 1982).

A.3
Multivariate Normal Distribution in Matrix Form

The multivariate normal distribution is very important in multivariate statistical applications. The distribution is given by (Marriot, 1974):

$$f(x_1, x_2, \ldots x_p) = K \exp\left[-\frac{1}{2} \sum_{r,s=1}^{P} \alpha_{rs}(x_r - \mu_r)(x_s - \mu_s) \right] \qquad (A.3.1)$$

and in matrix form:

$$f(X) = K \exp\left[-\frac{1}{2}(x - \mu)'\alpha(x - \mu)\right] \tag{A.3.2}$$

where μ represents the vector of means of p variates, and α is the matrix inverse to the dispersion matrix. The dispersion matrix is the covariance matrix with elements equal to the variances and covariances of the p variates. It is thus a $p \times p$ symmetric matrix with $1/2p\,(p + 1)$ distinct elements. The distribution also involves $1/2p\,(p + 3)$ parameters, made up of p variate means, p variances, and the $1/2p\,(p - 1)$ covariances (or correlations) between the variates. Since the correlations cannot exceed one, this implies that the variance-covariance matrix and its inverse must be positive definite (or positive semi-definite in the degenerate case when one of the variates can be expressed as a linear combination of the others). The multivariate central limit theorem is true for any distribution whatever with finite variance, but its practical importance is that it justifies the use of theory based on the normal distribution even when the distribution of x is not strictly normal, i.e., for slightly skewed distributions, for rectangular, binomial, and Poisson distributions, or other distributions, that do not have too long a tail (Marriott 1974). It does not work for observations with a few extreme outliers, because the normal distribution is approached too slowly, and the approximation is valid only for extremely large values of N. Thus these cases require a transformation. The central limit theorem is what gives the multivariate normal distribution its importance in multivariate statistics and ensures that many of the statistical techniques and tests based on multivariate normal distribution theory are robust, and will not give seriously misleading results even though the original data are not derived from a multivariate normal distribution.

The chi-square distribution is the basis of all univariate significance tests based on normal distribution and the multivariate generalization of this distribution is the Wishart distribution, the joint distribution of the sample variances and covariances (Wishart 1928).

References

Abrahams AD (1972) Factor analysis of drainage basin properties: evidence for stream abstraction accompanying the degradation of relief. Water Resour Res 8, 3:624-633

Afifi AA, Azen SP (1972) Statistical analysis - a computer oriented approach. Academic Press, New York, 366 pp

Afifi AA, Clark V (1990) Computer-aided multivariate analysis, 2nd edn. Van Nostrand Reinhold, New York, 505 pp

Agresti A (1990) Categorical data analysis. John Wiley, New York, 558 pp

Ahmed S, de Marsily G (1989) Comparison of geostatistical methods for estimating transmissivity and specific capacity. Water Resour Res 23:1717-1737

Ali JA (1984) Some trace-element analysis of Pliocene molasse in recent Euphrates and Tigris fluvial sediments. J Chem Geol 45:213-224

Anderson TW (1958) An introduction to multivariate statistical analysis. John Wiley, New York

Andrews DF (1972) Plots of high dimensional data. Biometrics 28:125-136

Arkin H, Colton R (1962) Tables for statisticians, College Outline Series. Barnes and Noble, New York, 152 pp

Armstrong M (ed) (1988) Geostatistics-Proc 3rd Int. Geostatistics Congr, Sept 5-9, 1988, Avignon, France. Kluwer, Boston, 1038 pp

Ashford JR, Sowden RR (1970) Multivariate probit analysis. Biometrics 26:535-545

Atkinson SE (1978) Small sample properties of simultaneous estimators with multicollinearity. J Am Statist Assoc 73, 364:719-723

Bloomfield JA (1976) The application of cluster analysis to stream water quality data. In: Proc Conf on Environmental Modeling and Simulation. US Environmental Protection Agency, Washington DC, pp 683-690

Box GEP, Draper NR (1965) The Bayesian estimation of common parameters from several responses. Biometrika 52:355-365

Box GEP, Jenkins GM (1970) Time series analysis: forecasting and control, 1st edn. Holden-Day, San Francisco

Bos GEP, Jenkins GM (1976) Time series analysis: forecasting and control, 2nd edn. Holden-Day, San Francisco

Box GEP, Lucas HL (1959) Design of experiments in nonlinear situations. Biometrika 46:77-90

Brown CE (1977) Multivariate analysis of petrographic and chemical properties influencing porosity and permeability in selected carbonate aquifers in central Pennsylvania. Doctorial Diss, The Pennsylvania State University, University Park, 206 pp

Brown CE, Mose DG , Mushrush GW, Chrosniak CE (1992) Statistical analysis of the radon-222 potential of rocks in Virginia, USA. J Environ Geol Water Sci 19, 3:193-203

Brown CE (1993) Use of principal component, correlation, and stepwise multiple regression analyses to investigate selected physical and hydraulic properties of carbonate rock aquifers. J Hydrol 147:169-195

Bulter JC (1974) Analysis of correlation between percentages. J Geol Educ March 1974:56-61

Carlson RF, MacCormick AJA, Watts DG (1970) Application of linear random models to four annual streamflow series. Water Resour Res 6, 4:1070-1078

Chan WYT, Wallis KF (1978) Multiple time series modelling, another look at the mink-muskrat interaction. Appl Statist 27:168-175

Chatfield C (1984) The analysis of time series data: an introduction. Chapman and Hall, New York, 286 pp

Chatfield C, Collins AJ (1980) Introduction to multivariate analysis. Chapman and Hall, New York, 246 pp

Chernoff H (1973) Using faces to represent points in k-dimensional space graphically. J Am Statist Assoc 68, 342:361–368

Chow VT, Kareliotis SJ (1970) Analysis of stochastic hydrologic systems. Water Resour Res 6 (6):1569–1582

Cochran JA (1960) The sampling problem in sedimentary petrography: a contribution. MS Thesis, The Pennsylvania State University, University Park, 94 pp

Cochran WG (1957) Analysis of covariance: its nature and uses. J Biometr 13, 3:261–281

Cooley WW, Lohnes PR (1971) Multivariate data analysis. John Wiley, New York, 364 pp

Curray JR, Griffiths JC (1955) Sphericity and roundness of quartz grains. Bull Geol Soc Am 66:1075–1096

Daniel WW, Terrell JC (1983) Business statistics. 3rd edn. Houghton Mifflin, Boston, 700 pp

Davis JC (1973) Statistics and data analysis in geology. John Wiley, New York, 550 pp

Davis JC (1986) Statistics and data analysis in geology. 2nd edn. John Wiley, New York, 646 pp

Dixon WJ (ed) (1974) BMD-Biomedical Computer Programs. University of California Press, Berkeley, 773 pp

Dixon DG, Sprague JB (1981) Acclimation to copper by rainbow trout (*Salmo gairdneri*) – a modifying factor in toxicity. Can J Fish Aquat Sci 38:880–888

Downing D, Clark J (1985) Business statistics. Barron's, New York, 466 pp

Drake JJ, Harmon RS (1973) Hydrochemical environments of carbonate terrains. Water Resour Res 9:949–957

Draper NR, Smith H (1981) Applied regression analysis. 2nd edn. John Wiley, New York 709 pp

Durbin J, Watson GS (1950) Testing for serial correlation 1. Biometrika 37:409–428

Durbin J, Watson GS (1951) Testing for serial correlation 2. Biometrika 38:159–178

Durbin J, Watson GS (1971) Testing for serial correlation 3. Biometrika 58:1–19

Eckhardt DA, Flipse WJ, Oaksford ET (1989) Relation between land use and ground-water quality in the upper glacial aquifer in Nassau and Suffolk counties, Long Island NY. US Geological Survey, WRI Rep 86–4142, GPO, Washington, DC, 26 pp

Ekwere SJ, Olade MA (1984) Geochemistry of the tinnobium bearing granites of the Liruei (Ririwai) complex, younger granite province, Nigeria. J Chem Geol 45:225–243

Erjavec J, Box GEP, Hunter WG, MacGregor JF (1973) Some problems associated with the analysis of multiresponse data. Technometrics:33–51

Fisher RA (1936) The use of multiple measurements in taxonomic problems. Ann Eugen 7:179–188

Fox WT (1975) Some practical aspects of time series analysis, Chap 3. In: McCammon, RB (ed) Concepts in geostatistics. Springer, Berlin Heidelberg New York, pp 71–89

Frost J, Clarke RT (1973) Use of cross correlation between hydrological time series to improve estimates of lag one autoregressive parameters. Water Resour Res 9:906

Gabriel RK (1978) A simple method of comparison of means. J Am Statist Assoc 73 (364):724–729

Gardner ES Jr (1985) Exponential smoothing: the state of the art. J Forecast 4:1–28

Gerrild PM, Lantz RJ (1969) Chemical analysis of 75 crude oil samples from pliocene sand units, Elk Hills Oil Field, California. US Geological Survey Open-File Rep 69–1277, Denver

Gilroy EJ (1970) Reliability of a variance estimate obtained from a sample augmented by multivariate regression. Water Resour Res 6 (6):1595–1600

Gittins R (1985) Canonical analysis: a review with applications in ecology. Springer, Berlin Heidelberg New York

Goodrich RL, Stellwagen EA (1991) Applied time series analysis and forecasting, vols 1 and 2, unpublished notes. The Institute for Professional Education, Arlington, Virginia

Griffiths JC (1955) Experimental designs in the earth sciences. Trans Am Geophys Union 36:1–11

Griffiths JC (1966) A genetic model for the interpretative petrology of detrital sediments. J Geol 74 (5) Part 2:653–671

Griffiths JC (1967) Scientific method in analysis of sediments. McGraw-Hill, New York, 508 pp

Haan CT, Allen DM (1972) Comparison of multiple regression and principal component regression for predicting water yields in Kentucky. Water Resour Res 8 (9):1593–1596

Helsel DR, Hirsch RM (1992) Statistical methods in water resources. Studies in environmental science, no 49. Elsevier, Now York, 522 pp

Hotelling H (1933) Analysis of a complex of statistical variables into principal components. J Educ Psych 24:417–441; 498–520

Hotelling H (1953) New light on the correlation coefficient and its transforms. J R Statist Soc 15:193–232

Houston JFT (1983) Ground-water systems simulation by time-series techniques. Ground Water 21 (3):301–310

Howarth RS, Sprague JB (1978) Copper lethality to rainbow trout in waters of various hardness and pH. Water Res 12:455–462

Hsu D-A, Hunter JS (1975) Time series analysis and forecasting for air pollution concentrations with seasonal variations. In: Proc Conf on Environmental Modeling and Simulation. US Environmental Protection Agency, Washington, DC, pp 673–677

Hull LC (1984) Geochemistry of ground-water in the Sacramento valley, CA. US Geological Survey Prof Pap 1401B, Government Printing Office, Washington, DC, pp B1–B36

Isaaks EH, Srivastava RM (1989) An introduction to applied geostatistics. Oxford University Press, New York, 561 pp

Jenkins GM (1982) Some practical aspects of forecasting in organizations. J Forecast 1:3–21

Johnson DE (1987) Applied multivariate methods using popular statistical computing packages. Unpublished notes: Institute of Professional Education, Arlington, Virginia, 251 pp

Johnston RA, Wichern DW (1988) Applied mltivariate statistical analysis. Prentice Hall, Englewood Cliffs, 607 pp

Jolliffe IT (1986) Principal component analysis. Springer, Berlin Heidelberg New York

Journel AG, Huijbregts CJ (1978) Mining geostatistics. Academic Press, New York, 600 pp

Kaiser HF (1974) An index of factorial simplicity. Psychometrika 39:31–36

Korin BP (1968) On the distribution of a statistic used for testing a covariance matrix. Biometrika 55:171–178

Kurskal JB (1964) Multidimensional scaling by optimizing goodness of fit to a nonmetric hypothesis. Psychometrika 29:1–27

Leenaers H, Okx JP, Burrough PA (1989) Co-kriging: an accurate and inexpensive means of mapping soil pollution by using elevation data. In: Armstrong M (ed) Geostatistics, vol 1. Kluwer, Boston, pp 371–382

Ljung GM, Box GEP (1978) On a measure of lack of fit in time series models. Biometrika 65:297–303

Lungu EM, Sefe FTK (1991) Stochastic analysis of monthly streamflows. J Hydrol 126:171–182

Makridakis S (1986) The art and science of forecasting – an assessment and future directions. Int J Forecast 2:15–39

Makridakis S, Andersen A, Carbone R, Fildes R, Hilbon M, Lewandowski R, Newton J, Parzen E, Winkler R (1982) The accuracy of extrapolation (time series) methods: results of a forecasting competition. J Forecast 1:111–153

Mantel N, Valand RS (1970) A technique for nonparametric multivariate analysis. Biometrics, Sept 1970:547–558

Mardia KV, Kent JT, Bibby JM (1979) Multivariate analysis. Academic Press, New York, 521 pp

Marsily G de (1986) Quantitative hydrogeology. Academic Press, San Diego, 440 pp

Matheron G (1971) The theory of regionalized variables and its applications. Cahiers du Centre de Morphologie Mathematique de Fontainebleau, No 5. Ecole des Mines, Paris, 211 pp

Matheron G (1976) A simple substitute for conditional expectation: disjunctive kriging. In: Guarascio M, Huijbregts DM (eds) Advances in geostatistics in the mining industry. NATO ASI Ser C, vol 24, Reidel, Dordrecht, pp 221–236

Marriott FHC (1974) The interpretation of multiple observartions. Academic Press, New York, 117 pp

McCammon RB (1975) Concepts in geostatistics. Springer, Berlin Heidelberg New York, 168 pp

McCuen RH, Rawls WJ, Whaley BL (1979) Comparative evaluation of statistical methods for water supply forecasting. Water Resour Bull 15 (4):935–947

Milliken GA, Johnson DE (1984) Analysis of messy Data, vol 1: Designed experiments. Van Nostrand Reinhold, New York, 473 pp

Milliken GA, Johnson DE (1989) Analysis of messy data, vol 3. Analysis of covariance. Kansas State Univ, Dept of Statistics Report

Minitab Inc (1986) Data analysis software reference manual. State College, Pennsylvania, 266 pp

Muge FH, Cabecadas G (1989) A geostatistical approach to eutrophication modelling. In: Armstrong M (ed) Geostatistics, vol 1. Kluwer, Boston, pp 445–457

Muirhead RJ, Waternaux CM (1980) Asymptotic distributions in canonical correlation analysis and other multivariate procedures for non-normal populations. Biometrika 67 (1): 31–43

Nagarsenker BN, Pillai KCS (1973) Distribution of the likelihood ratio criterion for testing a hypothesis specifying a covariance matrix. Biometrika 60 (2):359–364

Nie NH, Hull CH, Jenkins JG, Steinbrenner K, Bent DH (1975) SPSS – statistical package for the social sciences. McGraw-Hill, New York, 675 pp

Norusis MJ (1985) Advanced statistics. SPSS, Chicago, 505 pp

Oliver MA, Webster R (1986) Semi-variograms for modelling the spatial pattern of landform and soil properties. Earth Surface Processes Landforms 11:491–504

Ondrick CW, Srivastava GS (1970) Corfan – Fortran IV computer program for correlation, factor analysis (R- and Q-mode) and varimax rotation. Kansas Geological Survey Computer Contribution 42, Lawrence, Kansas, 92 pp

Pall GA (1971) Introduction to scientific computing. Meredith, New York, 677 pp

Parks JM (1966) Cluster analysis applied to multivariate geologic problems. J Geol 74:703–715

Pearson K (1901) On lines and planes of closet fit to systems of points in space. Philos Mag 2:559–572

Phillips RA (1988) Relationship between glacial geology and stream water chemistry in an area receiving acid deposition. Hydrol 101:263–273

Press J, Wilson S (1978) Choosing between logistics regression and discriminant analysis. J Am Statist Assoc 73 (364):699–706

Rand GM, Petrocelli SR (eds) (1985) Fundamentals of aquatic toxicology. Hemisphere Publishing (distributed by Taylor & Frances, Bristol, Pennsylvania), 666 pp

Ratha DS, Sahu BK (1993) Multivariate canonical correlation techniques: an economic approach for evaluation of pollutants in soil and sediments of Bombay region, India. J Environ Geol 21:201–207

Rauch RW, White WB (1970) Lithologic controls on development of solution porosity in carbonate aquifers. Water Resour Res 6 (4):1175–1192

Razack M, Dazy J (1990) Hydrochemical characterization of ground-water mixing in sedimentary and metamorphic reservoirs with combined use of Piper's principle and factor analysis. J Hydrol 114:371–393

Riley JA, Steinhorst RK, Winter GV, Williams RE (1990) Statistical analysis of the hydrochemistry of ground-waters in Columbia River Basalts. J Hydrol 119:245–262

Ruiz F, Gomis V, Blasco P (1990) Application of factor analaysis to hydrogeochemical study of a coastal aquifer. J Hydrol 119:169–177

SAS Institute, Inc (1985) SAS user's guide: statistics, version 5. Cary, North Carolina, 956 pp

Scanlon BR (1990) Relationship between ground-water contamination and major-ion chemistry in a karst aquifer. J Hydrol 119:271–291

Scheffe H (1958) Fitting straight lines when one variable is controlled. J Am Statist Assoc 53:106–118

Shepard RN (1980) Multidimensional scaling, tree-fitting, and clustering. Science 210, 4468:390–398

Siddiqui SH (1969) Hydrogeologic factors influencing well yields and aquifer hydraulic properties of folded and faulted carbonate rocks in central Pennsylvania. Doctoral Diss, The Pennsylvania State University, University Park, 502 pp

Smith HF (1957) Interpretation of adjusted treatment means and regressions in analysis of covariance. J Biometr 13 (3):282–309

Smyth JD, Istok JD (1989) Multivariate geostatistical analysis of ground-water contamination by pesticide and nitrate: a case history. In: Armstrong M (ed) Geostatistics, vol 2, Kluwer, Boston, pp 713–724

Snapinn SM, Small RD (1986) Tests of significance using regression models for ordered categorical data. Biometrics 42:583–592

Snedecor GW, Cochran WG (1967) Statistical methods, 6th ed. The Iowa State University Press, Ames

Spiegel MR (1961) Theory and problems of statistics. Schaum, New York, 359 pp

Steinhorst RK, Williams RE (1985) Discrimination of ground-water sources using cluster analysis, MANOVA, canonical analysis, and discriminant analysis. Water Resour Res 21 (8):1149–1156

Stevens SS (1946) On the theory of scales of measurement. Science 103:677–680

Tabachnick BG, Fidell LS (1989) Using multivariate statistics. Harper and Row, San Francisco, 746 pp

Tasker GD (1972) Estimating low-flow characteristics of streams in southeastern Massachusetts from maps of ground-water availability. US Geol Survey Prof Pap 800-D, US Government Printing Office, Washington, DC, pp d217–d220

Tasker GD, Stedinger JF (1989) An operational GLS model for hydrologic regression. J Hydrol 111:361–375

Theis CV (1935) The relationship between the lowering of the piezometric surface and the rate and duration of discharge of a well using ground-water storage. Trans Am Geophys Union 16:519–524

Tiao GC, Box GEP (1981) Modelling multivariate time series with applications. J Am Statist Assoc 76:802–816

Usunoff EJ, Guzman-Guzman A (1989) Multivariate analysis in hydrochemistry: an example of the use of factor and correspondence analyses. Ground Water 27, 1:27–34

Wackernagel H (1988) Geostatistical techniques for interpreting multivariate spatial information. In: Chung AG, Fabbri G, Sinding-Larsen R (eds) Quantitative analysis of mineral and energy resources. Reidel, Dordrecht, pp 393–409

Wackernagel H, Webster R, Oliver MA (1988) A geostatistical method for segmenting multivariate sequences of soil data. In: Bock HH (ed) Classification and related methods of data analysis. Elsevier – North-Holland, Amsterdam, pp 641–650

Wackernagel H, Petitgas P, Touffait Y (1989) Overview of methods for co-regionalization. In: Armstrong M (ed) Geostatistics, vol 1. Kluwer, Boston, pp 409–420

Wang HF, Anderson MP (1982) Introduction to ground-water modelling – finite difference and finite element methods. WH Freeman, San Francisco, 237 pp

Webster R (1978) Optimally partitioning soil transects. J Soil Sci 29:388–402

Webster R, Oliver MA (1989) Disjunctive kriging in agriculture. In: Armstrong M (ed) Geostatistics, vol 1, Kluwer, Boston, pp 421–432

Williams BC (1991) Statistical and geochemical analysis of vadose and saturated pore waters in sulfidic mine waste tailings. In: Proc 4th Int Mineral Water Assoc Congr, Pörtschach, Austria, pp 305–316

Williams F (1979) Reasoning with statistics, 2nd edn. Holt, Rinehart & Winston, New York, 204 pp

Williams RE (1982) Statistical identification of hydraulic connections between the surface of a mountain and internal mineralized sources. Ground Water 20 (4):466–478

Wishart J (1928) The generalized product-moment distribution in samples from a multivariate normal population. Biometrika 20A:32–52

Author Index

Subject Index

Springer
and the
environment

At Springer we firmly believe that an
international science publisher has a
special obligation to the environment,
and our corporate policies consistently
reflect this conviction.
We also expect our business partners –
paper mills, printers, packaging
manufacturers, etc. – to commit
themselves to using materials and
production processes that do not harm
the environment. The paper in this
book is made from low- or no-chlorine
pulp and is acid free, in conformance
with international standards for paper
permanency.

 Springer

Printing: Saladruck, Berlin
Binding: Buchbinderei Lüderitz & Bauer, Berlin